# Computer Explorations in Signals and Systems

## Using MATLAB®

## Second Edition

## PRENTICE HALL SIGNAL PROCESSING SERIES

Alan V. Oppenheim, *Series Editor*

BRACEWELL *Two Dimensional Imaging*
BRIGHAM *The Fast Fourier Transform and Its Applications (AOD)*
BUCK, DANIEL & SINGER *Computer Explorations in Signals and Systems Using MATLAB, 2/E*
CASTLEMAN *Digital Image Processing*
COHEN *Time-Frequency Analysis*
CROCHIERE & RABINER *Multirange Digital Signal Processing (AOD)*
JOHNSON & DUDGEON *Array Signal Processing (AOD)*
KAY *Fundamentals of Statistical Signal Processing, Vols. I and II*
KAY *Modern Spectral Estimation (AOD)*
LIM *Two-Dimensional Signal and Image Processing*
MCCLELLAN et al, *Computer-Based Exercises in Signal Processing using MATLAB5*
MENDEL *Lessons in Estimation Theory for Signal Processing Communications and Control 2/E*
NIKIAS & PETROPULU *Higher Order Spectra Analysis*
OPPENHEIM & SCHAFER *Digital Signal Processing*
OPPENHEIM & SCHAFER *Discrete-Time Signal Processing*
OPPENHEIM WILLSKY, WITH NAWAB *Signals and Systems*
ORFANIDIS *Introduction to Signal Processing*
PHILLIPS & NAGLE *Digital Control Systems Analysis and Design, 3/E*
QUATIERI *Discrete-Time Speech Signal Processing: Principles and Practice*
RABINER & JUANG *Fundamentals of Speech Recognition*
RABINER & SCHAFER *Digital Processing of Speech Signals*
STEARNS & DAVID *Signal Processing Algorithms in MATLAB*
TEKLAP *Digital Video Processing*
VAIDYANATHAN *Multirate Systems and Filter Banks*
VETTERLI & KOVACEVIC *Wavelets and Subband Coding*
WANG, *Video Processing and Communications*
WIDROW & STEARNS *Adaptive Signal Processing*

# Computer Explorations in Signals and Systems Using MATLAB®
## Second Edition

JOHN R. BUCK
*Department of Electrical and Computer Engineering*
*University of Massachusetts Dartmouth*

MICHAEL M. DANIEL
*Fidelity Investments*
*Cambridge, MA*

ANDREW C. SINGER
*Department of Electrical and Computer Engineering*
*University of Illinois at Urbana-Champaign*

*PRENTICE HALL SIGNAL PROCESSING SERIES*

PRENTICE HALL
Upper Saddle River, New Jersey 07458

**Library of Congress Cataloging-in-Publication Data**

CIP data on file.

Vice President and Editorial Director, ECS: *Marcia Horton*
Publisher: *Tom Robbins*
Associate Editor: *Alice Dworkin*
Editorial Assistant: *Jody McDonnell*
Vice President and Director of Production and Manufacturing, ESM: *David W. Riccardi*
Executive Managing Editor: *Vince O'Brien*
Managing Editor: *David A. George*
Production Editor: *Barbara A. Till*
Director of Creative Services: *Paul Belfanti*
Creative Director: *Carole Anson*
Art Director: *Jayne Conte*
Art Editor: *Greg Dulles*
Cover Designer: *Bruce Kenselaar*
Manufacturing Manager: *Trudy Pisciotti*
Manufacturing Buyer: *Lisa McDowell*
Marketing Manager: *Holly Stark*
Marketing Assistant: *Karen Moon*
Composition: *PreTEX, Inc.*

The author and publisher of this book have used their best efforts in preparing this book. These efforts include the development, research, and testing of the theories and programs to determine their effectiveness. The author and publisher make no warranty of any kind, expressed or implied, with regard to these programs or the documentation contained in this book. The author and publisher shall not be liable in any event for incidental or consequential damages in connection with, or arising out of, the furnishing, performance, or use of these programs.



Printed in the United States of America

10 9 8 7 6 5 4 3

ISBN 0-13-042155-3

Pearson Education Ltd., *London*
Pearson Education Australia Pty. Ltd., *Sydney*
Pearson Education Singapore, Pte. Ltd.
Pearson Education North Asia Ltd., *Hong Kong*
Pearson Education Canada, Inc., *Toronto*
Pearson Educacìon de Mexico, S.A. de C.V.
Pearson Education—Japan, *Tokyo*
Pearson Education Malaysia, Pte. Ltd.

To Our Families

# Contents

# Preface

This book provides computer exercises for an undergraduate course on signals and linear systems. Such a course or sequence of courses forms an important part of most engineering curricula. This book was primarily designed as a companion to the second edition of *Signals and Systems* by Oppenheim and Willsky with Nawab. While the sequence of chapter topics and the notation of this book match that of *Signals and Systems*, this book of exercises is self-contained and the coverage of fundamental theory and applications is sufficiently broad to make it an ideal companion to any introductory signals and systems text or course.

We believe that assignments of computer exercises in parallel with traditional written problems can help readers to develop a stronger intuition and a deeper understanding of signals and linear systems. To this end, the exercises require the readers to compare the answers they compute in MATLAB® with results and predictions made based on their analytic understanding of the material. The second edition has been updated to MATLAB 6, Release 12. We believe this approach actively challenges and involves the reader, providing more benefit than a passive computer demonstration. Wherever possible, the exercises have been divided into Basic, Intermediate, and Advanced Problems. In working the problems, the reader progresses from fundamental theory to real applications such as speech processing, financial market analysis and designing mechanical or communication systems. Basic Problems provide detailed instructions for readers, guiding them through the issues explored, but still requiring a justification of their results. Intermediate Problems examine more sophisticated concepts, and demand more initiative from the readers in their use of MATLAB. Finally, Advanced Problems challenge the readers' understanding of the more subtle or complicated issues, often requiring open-ended work, writing functions, or processing real data. Some of the Advanced Problems in this category are appropriate for advanced undergraduate coursework on signals and systems.

Care has been taken to ensure that all the exercises in this book can be completed within MATLAB 6. To assist readers, a list of MATLAB functions used in the text can be found in the index, which notes the exercise or page number in which they are explained. Throughout this book, MATLAB functions, commands, and variables will be indicated by `typewriter font`. The Ⓢ symbol following the title of an exercise indicates that the exercise requires the Symbolic Math Toolbox.

A number of exercises refer to functions or data files the reader will need. These are in the Computer Explorations Toolbox, which is available from the MathWorks, Inc. via the World Wide Web. Contact the MathWorks, Inc. about these files at:

The MathWorks, Inc.
3 Apple Hill Drive
Natick, MA 01760-2098
Phone: (508) 647-7000
Fax: (508) 647-7001
E-mail: info@mathworks.com
WWW: http://www.mathworks.com
ftp://ftp.mathworks.com/pub/books/buck

We would like to acknowledge and thank Alan Oppenheim and Alan Willsky for their support and encouragement during this project. They generously provided us with the opportunity to write this book, and then graciously and trustingly gave us space to pursue it independently. We would also like to thank the friends and colleagues at MIT with whom we have taught and worked over the years, especially Steven Isabelle, Hamid Nawab, Jim Preisig, Stephen Scherock, and Kathleen Wage. This book has certainly benefited from our interactions with them, and they are responsible for none of its shortcomings. Thanks also to Mukaya Panich and Krishna Pandey for diligently testing the exercises. Naomi Fernandes at the MathWorks, Inc. provided welcome assistance in setting up the internet site. The patience and support of Prentice Hall, especially Alice Dworkin, Marcia Horton, and Tom Robbins, has been instrumental in completing this project. We are also especially indebted to Peter Shargo and Farinaz Edalat at the University of Illinois for their help revising the book and testing all of the exercises with MATLAB 6 for the second edition.

JOHN R. BUCK
*Department of Electrical and Computer Engineering*
*University of Massachusetts Dartmouth*

MICHAEL M. DANIEL
*Fidelity Investments*
*Cambridge, MA*

ANDREW C. SINGER
*Department of Electrical and Computer Engineering*
*University of Illinois at Urbana-Champaign*

# Computer Explorations in Signals and Systems Using MATLAB®
## Second Edition

Chapter 1

# Signals and Systems

The basic concepts of signals and systems arise in a variety of contexts, from engineering design to financial analysis. In this chapter, you will learn how to represent, manipulate, and analyze basic signals and systems in MATLAB. The first section of this chapter, Tutorial ??, covers some of the fundamental tools used to construct signals in MATLAB. This tutorial is meant to be a supplement to, but not a substitute for, the tutorial given in *The MATLAB User's Guide.* If you have not already done so, you are strongly encouraged to work through this tutorial or the demos found in the MATLAB software (e.g. `intro`) before beginning this chapter. While not all of the MATLAB functions introduced in these tutorials are needed for the exercises of this chapter, most will be used at some point in this book.
Complex exponential signals are used frequently for signals and systems analysis, in part because complex exponential signals form the building blocks of large classes of signals. Exercise ?? covers the MATLAB functions required for generating and plotting discrete-time sinusoidal signals, which are equal to the sum of two discrete-time complex exponential signals, i.e.,

$$\cos(\omega n) = \frac{1}{2}\left(e^{i\omega n} + e^{-i\omega n}\right), \tag{1.1}$$

$$\sin(\omega n) = \frac{1}{2i}\left(e^{i\omega n} - e^{-i\omega n}\right). \tag{1.2}$$

Exercise ?? shows how to plot discrete-time signals $x[n]$ after transformations of the independent variable $n$. The next two exercises cover system representations in MATLAB. For Exercise ??, you must demonstrate your understanding of basic system properties like linearity and time-invariance. For Exercise ??, you must implement a system described by a first-order difference equation.
Several of the exercises in this chapter use the Symbolic Math Toolbox to study basic signals and systems. In Exercise ??, you will construct symbolic expressions for continuous-time complex exponential signals, which have the form $e^{st}$ for some complex number $s$. (Note that both $i$ and $j$ will be used in this book to represent the imaginary number $\sqrt{-1}$.) Exercise ?? uses the Symbolic Math Toolbox to implement transformations on the time-index of continuous-time signals. For Exercise ??, you must create analytic expressions for the energy of periodic signals and relate energy to time-averaged power.

## 1.1 Tutorial: Basic MATLAB Functions for Representing Signals

In this tutorial, you will learn how to use several MATLAB functions that will frequently be used to construct and manipulate signals in this book. If you have not already done so, you are strongly encouraged to work through the tutorials in *The MATLAB User's Guide* or in the MATLAB software. This tutorial is not meant to replace those tutorials, but rather to illustrate how some of the functions described there can be used for representing and working with signals. Although there are no problems to be worked in this tutorial, you should duplicate all the examples in MATLAB to give yourself practice with the commands. In general, signals will be represented by a row or column vector, depending on the context. All vectors represented in MATLAB are indexed starting with 1, i.e., `y(1)` is the first element of the vector `y`. If these indices do not correspond to those in your application, you can create an additional index vector to properly keep track of the signal index. For example, to represent the discrete-time signal

$$x[n] = \begin{cases} 2n\,, & -3 \leq n \leq 3\,, \\ 0\,, & \text{otherwise}\,, \end{cases}$$

you could first use the colon operator to define the index vector for the nonzero samples of $x[n]$, and then define the vector `x` to contain the values of the signal at each of these time indices:

```
>> n = [-3:3];
>> x = 2*n;
```

Note that we have used semicolons at the end of each command to suppress unnecessary MATLAB echoing. For instance, without the semicolon you would get

```
>> n = [-3:3]
n =
    -3    -2    -1     0     1     2     3
```

You can plot this signal by typing `stem(n,x)`. If you want to examine the signal over a wider range of indices, you will need to extend both `n` and `x`. For instance, if you want to plot the signal over the range $-5 \leq n \leq 5$, you can extend the index vector `n`, then add additional elements to `x` for these new samples:

```
>> n = [-5:5];
>> x = [0 0 x 0 0 ];
```

If you want to greatly extend the range of the signal, you may find it helpful to use the function `zeros`. For instance, if you wanted to include the region $-100 \leq n \leq 100$, after you had already extended `x` to include $-5 \leq n \leq 5$ as shown above, you could type

```
>> n = [-100:100];
>> x = [zeros(1,95) x zeros(1,95)];
```

Suppose you want to define $x_1[n]$ to be the discrete-time unit impulse function and $x_2[n]$ to be a time-advanced version of $x_1[n]$, i.e., $x_1[n] = \delta[n]$ and $x_2[n] = \delta[n+2]$. You could represent these signals in MATLAB by typing

```
>> nx1 = [0:10];
>> x1 = [1 zeros(1,10)];
>> nx2 = [-5:5];
>> x2 = [zeros(1,3) 1 zeros(1,7)];
```

You could then plot these signals by `stem(nx1,x1)` and `stem(nx2,x2)`. If you did not define the index vectors, and simply typed `stem(x1)` and `stem(x2)`, you would make plots of the signals $\delta[n-1]$ and $\delta[n-4]$ and not of the desired signals. The index vector will also be useful for keeping track of the time origin of a vector when you work on more advanced exercises in later chapters.

We will explore two methods for representing continuous-time signals in MATLAB. One method is to use the Symbolic Math Toolbox. The exercises in this book which use the Symbolic Math Toolbox are marked by the symbol Ⓢ at the end of the exercise title. You can also represent continuous-time signals with vectors containing closely spaced samples of the signals in time. The projects in early chapters that represent continuous-time signals by closely spaced samples will always explicitly specify the time spacing to use to guarantee that the signal is accurately represented. In Chapter **??**, you will explore the issues involved with representing a continuous-time signal by discrete-time samples. Vectors of closely spaced time indices can be created in a number of ways. Two simple methods are to use the colon operator with the optional step argument, and to use the `linspace` function. For instance, if you wanted to create a vector that covered the interval $-5 \leq t \leq 5$ in steps of 0.1 seconds, you could either use `t=[-5:0.1:5]` or `t=linspace(-5,5,101)`.

Sinusoids and complex exponentials are important signals for the study of linear systems. MATLAB provides several functions that are useful for defining such signals, especially if you have already defined either a continuous-time or discrete-time index vector. For instance, if you wanted to form a vector to represent $x(t) = \sin(\pi t/4)$ for $-5 \leq t \leq 5$, you could use the vector `t` defined in the previous paragraph and type `x=sin(pi*t/4)`. Note that when the argument to `sin` (or many other MATLAB functions, such as `cos` and `exp`) is a vector, the function returns a vector of the same size where each element of the output vector is the function applied to the corresponding element of the input vector. You can use the `plot` command to plot your approximation to the continuous-time signal $x(t)$. Unlike `stem`, `plot` connects adjacent elements with a straight line, so that when the time index is finely sampled, the straight lines are a close approximation to a plot of the original continuous-time signal. For this example, you can generate such a plot by typing `plot(t,x)`. In general, you will want to use `stem` to plot short discrete-time sequences, and `plot` for sampled approximations of continuous-time signals or for very long discrete-time signals where the number of stems grows unwieldy.

Discrete-time sinusoids and complex exponentials can also be generated using `cos`, `sin`, and `exp`. For instance, to represent the discrete-time signal $x[n] = e^{j(\pi/8)n}$ for $0 \leq n \leq 32$, you would type

```
>> n = [0:32];
>> x = exp(j*(pi/8)*n);
```

The vector x now contains the complex values of the signal $x[n]$ over the interval $0 \leq n \leq 32$. To plot complex signals, you must plot their real and imaginary parts, or magnitude and angle, separately. The MATLAB functions `real`, `imag`, `abs`, and `angle` compute these functions of a complex vector on a term-by-term basis. You can plot each of these functions of this complex signal by typing

```
>> stem(n,real(x))
>> stem(n,imag(x))
>> stem(n,abs(x))
>> stem(n,angle(x))
```

For the last example, note that the value returned by `angle` is the phase of the complex number in radians. To convert to degrees, type `stem(n,angle(x)*(180/pi))`.
MATLAB also allows you to add, subtract, multiply, divide, scale and exponentiate signals. As long as the vectors representing the signals have the same time-origins and the same number of elements, e.g.,

```
>> x1 = sin((pi/4)*[0:15]);
>> x2 = cos((pi/7)*[0:15]);
```

you can perform the following term-by-term operations:

```
>> y1 = x1+x2;
>> y2 = x1-x2;
>> y3 = x1.*x2;
>> y4 = x1./x2;
>> y5 = 2*x1;
>> y6 = x1.^3;
```

Note that for multiplying, dividing and exponentiating on a term-by-term basis, you must precede the operator with a period, i.e., use the `.*` function instead of just `*` for term-by-term multiplication. MATLAB interprets the `*` operator without a period to be the matrix multiplication operator, not term-by-term multiplication. For example, if you try to multiply x1 and x2 using `*`, you will receive the following error message:

```
>> x1*x2
???  Error using ==> *
Inner matrix dimensions must agree.
```

because matrix multiplication requires that the number of columns of the first argument be equal to the number of rows of the second argument, which is not true for the two $1 \times 5$ vectors x1 and x2. You must also be careful to use `./` and `.^` when operating on vectors

term-by-term, since / and ^ are matrix operations.
MATLAB also includes several commands to help you label plots appropriately, as well as to print them out. The **title** command places its argument over the current plot as the title. The commands **xlabel** and **ylabel** allow you to label the axes of your graph, making it clear what has been plotted. Every plot or graph you generate should have a title, as well as labels for both axes. For example, consider again a plot of the following signal and index vector:

```
>> n = [0:32];
>> x = exp(j*(pi/8)*n);
>> stem(n,angle(x))
```

You could label your graph by typing

```
>> title('Phase of exp(j*(pi/8)*n)')
>> xlabel('n (samples)')
>> ylabel('Phase of x[n] (radians)')
```

The **print** command allows you to print out the current plot. You should type **help print** to understand how it works on your system, as it will vary slightly depending on the operating system and configuration of the computer you are using.
Another important feature of MATLAB is the ability to write M-files. There are two types of M-files: functions and command scripts . A command script is a text file of MATLAB commands whose filename ends in `.m` in the current working directory or elsewhere on your MATLABPATH. If you type the name of this file (without the `.m`), the commands contained in the file will be executed. Using these scripts will make it much easier for you to do the exercises in this book. Many exercises will require you to process several signals in a similar or identical way. If you do not use scripts, you will have to retype all the commands anew. However, if you did the first problem using a script, you can process all the subsequent signals in that exercise by copying the script file and editing it to process the new signal. For example, suppose you had the following script file `prob1.m` to plot the discrete-time signal $\cos(\pi n/4)$ and compute its mean over the interval $0 \leq n \leq 16$:

```
% prob1.m
n = [0:16];
x1 = cos(pi*n/4);
y1 = mean(x1);
stem(n,x1)
title('x1 = cos(pi*n/4)')
xlabel('n (samples)')
ylabel('x1[n]')
```

If you then wanted to do the same for $x_2[n] = \sin(\pi n/4)$, you could copy `prob1.m` to `prob2.m`, then edit it slightly to get

```
% prob2.m
n = [0:16];
x2 = sin(pi*n/4);
y2 = mean(x2);
stem(n,x2)
title('x = sin(pi*n/4)')
xlabel('n (samples)')
ylabel('x2[n]')
```

You can then type `prob2` to run these commands and generate the desired plot and compute the average value of the new signal. Instead of retyping all 7 lines, you need only edit about a dozen characters. We strongly encourage you to use scripts in working the problems in this book, with a separate script for each exercise, or even each problem. Scripts also make debugging your work much easier, as you can fix one mistake and then easily rerun the modified sequence of commands. Finally, when you complete an exercise, it is easy to print out your script file and hand it in as a record of your work.
An M-file implementing a function is a text file with a title ending in `.m` whose first word is `function`. The rest of the first line of the file specifies the names of the input and output arguments of the function. For example, the following M-file implements a function called `foo` which accepts an input `x` and returns `y` and `z`, which are equal to `2*x` and `(5/9)*(x-32)`, respectively:

```
function [y,z] = foo(x)
% [y,z] = foo(x) accepts a numerical argument x and
% returns two arguments y and z, where y is 2*x
% and z is (5/9)*(x-32)
y = 2*x;
z = (5/9)*(x-32);
```

Two sample calls to `foo` are shown below:

```
>> [y,z] = foo(-40)
y =
   -80
z =
   -40

>> [y,z] = foo(212)
y =
   424
z =
   100
```

The commands described in this tutorial are by no means the complete set you will need to do the exercises in this book, but instead are meant to get you started using MATLAB. Future exercises in this book will assume that you are comfortable using the commands discussed here, and that you are also able to learn about other basic mathematical commands in MATLAB by using either the manual or the `help` function. Specialized functions for signal processing will often be described in their own tutorials in later chapters. Again, if you have not already done so, you should work through the general tutorial in the MATLAB manual so you are familiar with the functions available in MATLAB.

## 1.2 Discrete-Time Sinusoidal Signals

Discrete-time complex exponentials play an important role in the analysis of discrete-time signals and systems. A discrete-time complex exponential has the form $\alpha^n$, where $\alpha$ is a complex scalar. The discrete-time sine and cosine signals can be built from complex exponential signals by setting $\alpha = e^{\pm i\omega}$, namely,

$$\cos(\omega n) = \frac{1}{2}\left(e^{i\omega n} + e^{-i\omega n}\right), \tag{1.3}$$

$$\sin(\omega n) = \frac{1}{2i}\left(e^{i\omega n} - e^{-i\omega n}\right). \tag{1.4}$$

In this exercise, you will create and analyze a number of discrete-time sinusoids. There are many similarities between continuous-time and discrete-time sinusoids, as follows from a simple comparison of Eqs. (??)-(??) and Eqs. (??)-(??). However, you will also examine some of the important differences between sinusoids in continuous and discrete time in this exercise.

### Basic Problems

(a). Consider the discrete-time signal

$$x_M[n] = \sin\left(\frac{2\pi Mn}{N}\right),$$

and assume $N = 12$. For $M = 4$, 5, 7, and 10, plot $x_M[n]$ on the interval $0 \le n \le 2N-1$. Use `stem` to create your plots, and be sure to appropriately label your axes. What is the fundamental period of each signal? In general, how can the fundamental period be determined from arbitrary integer values of $M$ and $N$? Be sure to consider the case in which $M > N$.

(b). Consider the signal

$$x_k[n] = \sin(\omega_k n),$$

where $\omega_k = 2\pi k/5$. For $x_k[n]$ given by $k = 1$, 2, 4, and 6, use `stem` to plot each signal on the interval $0 \le n \le 9$. All of the signals should be plotted with separate axes in the same figure using `subplot`. How many unique signals have you plotted? If two signals are identical, explain how different values of $\omega_k$ can yield the same signal.

(c). Now consider the following three signals:

$$x_1[n] = \cos\left(\frac{2\pi n}{N}\right) + 2\cos\left(\frac{3\pi n}{N}\right),$$
$$x_2[n] = 2\cos\left(\frac{2n}{N}\right) + \cos\left(\frac{3n}{N}\right),$$
$$x_3[n] = \cos\left(\frac{2\pi n}{N}\right) + 3\sin\left(\frac{5\pi n}{2N}\right).$$

Assume $N = 6$ for each signal. Determine whether or not each signal is periodic. If a signal is periodic, plot the signal for two periods, starting at $n = 0$. If the signal is not periodic, plot the signal for $0 \le n \le 4N$ and explain why it is not periodic. Remember to use `stem` and to appropriately label your axes.

### Intermediate Problems

(d). Plot each of the following signals on the interval $0 \le n \le 31$:

$$x_1[n] = \sin\left(\frac{\pi n}{4}\right)\cos\left(\frac{\pi n}{4}\right),$$
$$x_2[n] = \cos^2\left(\frac{\pi n}{4}\right),$$
$$x_3[n] = \sin\left(\frac{\pi n}{4}\right)\cos\left(\frac{\pi n}{8}\right).$$

What is the fundamental period of each signal? For each of these three signals, how could you have determined the fundamental period without relying upon MATLAB?

(e). Consider the signals you plotted in Parts **??** and **??**. Is the addition of two periodic signals necessarily periodic? Is the multiplication of two periodic signals necessarily periodic? Clearly explain your answers.

## 1.3 Transformations of the Time Index for Discrete-Time Signals

In this exercise you will examine how to use MATLAB to represent discrete-time signals. In addition, you will explore the effect of simple transformations of the independent variable, such as delaying the signal or reversing its time axis. These rudimentary transformations of the independent variable will occur frequently in studying signals and systems, so becoming comfortable and confident with them now will benefit you in studying more advanced topics.

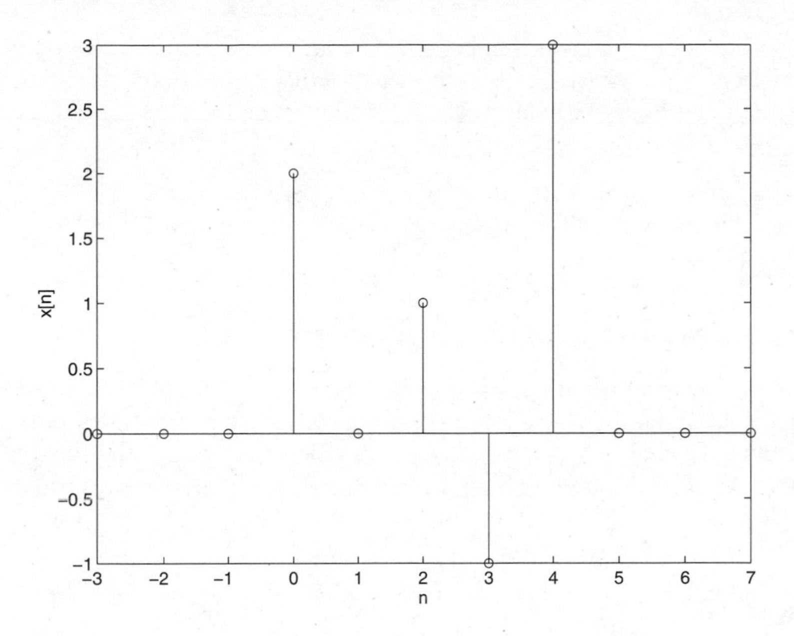


**Figure 1.1.** Discrete-time signal $x[n]$.

## Basic Problems

(a). Define a MATLAB vector `nx` to be the time indices $-3 \leq n \leq 7$ and the MATLAB vector `x` to be the values of the signal $x[n]$ at those samples, where $x[n]$ is given by

$$x[n] = \begin{cases} 2\,, & n = 0\,, \\ 1\,, & n = 2\,, \\ 1\,, & n = 3\,, \\ 3\,, & n = 4\,, \\ 0\,, & \text{otherwise}\,. \end{cases}$$

If you have defined these vectors correctly, you should be able to plot this discrete-time sequence by typing `stem(nx,x)`. The resulting plot should match the plot shown in Figure ??.

(b). For this part, you will define MATLAB vectors `y1` through `y4` to represent the following discrete-time signals:

$$\begin{aligned} y_1[n] &= x[n-2]\,, \\ y_2[n] &= x[n+1]\,, \\ y_3[n] &= x[-n]\,, \\ y_4[n] &= x[-n+1]\,. \end{aligned}$$

To do this, you should define `y1` through `y4` to be equal to `x`. The key is to define correctly the corresponding index vectors `ny1` through `ny4`. First, you should figure out how the index of a given sample of $x[n]$ changes when transforming to $y_i[n]$. The index vectors need not span the same set of indices as nx, but they should all be at least 11 samples long and include the indices of all nonzero samples of the associated signal.

(c). Generate plots of $y_1[n]$ through $y_4[n]$ using stem. Based on your plots, state how each signal is related to the original $x[n]$, e.g., "delayed by 4" or "flipped and then advanced by 3."

## 1.4 Properties of Discrete-Time Systems

Discrete-time systems are often characterized in terms of a number of properties such as linearity, time invariance, stability, causality, and invertibility. It is important to understand how to demonstrate when a system does or does not satisfy a given property. MATLAB can be used to construct counter-examples demonstrating that certain properties are not satisfied. In this exercise, you will obtain practice using MATLAB to construct such counter-examples for a variety of systems and properties.

### Basic Problems

For these problems, you are told which property a given system does not satisfy, and the input sequence or sequences that demonstrate clearly how the system violates the property. For each system, define MATLAB vectors representing the input(s) and output(s). Then, make plots of these signals, and construct a well reasoned argument explaining how these figures demonstrate that the system fails to satisfy the property in question.

(a). The system $y[n] = \sin((\pi/2)x[n])$ is not linear. Use the signals $x_1[n] = \delta[n]$ and $x_2[n] = 2\delta[n]$ to demonstrate how the system violates linearity.

(b). The system $y[n] = x[n] + x[n+1]$ is not causal. Use the signal $x[n] = u[n]$ to demonstrate this. Define the MATLAB vectors x and y to represent the input on the interval $-5 \leq n \leq 9$, and the output on the interval $-6 \leq n \leq 9$, respectively.

### Intermediate Problems

For these problems, you will be given a system and a property that the system does not satisfy, but must discover for yourself an input or pair of input signals to base your argument upon. Again, create MATLAB vectors to represent the inputs and outputs of the system and generate appropriate plots with these vectors. Use your plots to make a clear and concise argument about why the system does not satisfy the specified property.

(c). The system $y[n] = \log(x[n])$ is not stable.

(d). The system given in Part ?? is not invertible.

### Advanced Problems

For each of the following systems, state whether or not the system is linear, time-invariant, causal, stable, and invertible. For each property you claim the system does not possess, construct a counter-argument using MATLAB to demonstrate how the system violates the property in question.

(e). $y[n] = x^3[n]$

(f). $y[n] = n\,x[n]$

(g). $y[n] = x[2n]$.

## 1.5 Implementing a First-Order Difference Equation

Discrete-time systems are often implemented with linear constant-coefficient difference equations. Two very simple difference equations are the first-order moving average

$$y[n] = x[n] + b\,x[n-1]\,, \tag{1.5}$$

and the first-order autoregression

$$y[n] = a\,y[n-1] + x[n]\,. \tag{1.6}$$

Even these simple systems can be used to model or approximate a number of practical systems. For instance, the first-order autoregression can be used to model a bank account, where $y[n]$ is the balance at time $n$, $x[n]$ is the deposit or withdrawal at time $n$, and $a = 1+r$ is the compounding due to interest rate $r$. In this exercise, you will be asked to write a function which implements the first-order autoregression equation. You will then be asked to test and analyze your function on some example systems.

### Advanced Problems

(a). Write a function `y=diffeqn(a,x,yn1)` which computes the output $y[n]$ of the causal system determined by Eq. (??). The input vector `x` contains $x[n]$ for $0 \le n \le N-1$ and `yn1` supplies the value of $y[-1]$. The output vector `y` contains $y[n]$ for $0 \le n \le N-1$. The first line of your M-file should read

```
function y = diffeqn(a,x,yn1)
```

Hint: Note that $y[-1]$ is necessary for computing $y[0]$, which is the first step of the autoregression. Use a `for` loop in your M-file to compute $y[n]$ for successively larger values of $n$, starting with $n = 0$.

(b). Assume that $a = 1$, $y[-1] = 0$, and that we are only interested in the output over the interval $0 \le n \le 30$. Use your function to compute the response due to $x_1[n] = \delta[n]$ and $x_2[n] = u[n]$, the unit impulse and unit step, respectively. Plot each response using `stem`.

(c). Assume again that $a = 1$, but that $y[-1] = -1$. Use your function to compute $y[n]$ over $0 \le n \le 30$ when the inputs are $x_1[n] = u[n]$ and $x_2[n] = 2\,u[n]$. Define the outputs produced by the two signals to be $y_1[n]$ and $y_2[n]$, respectively. Use `stem` to display both outputs. Use `stem` to plot $(2\,y_1[n] - y_2[n])$. Given that Eq. (??) is a linear difference equation, why isn't this difference identically zero?

(d). The causal systems described by Eq. (??) are BIBO (bounded-input bounded-output) stable whenever $|a| < 1$. A property of these stable systems is that the effect of the initial condition becomes insignificant for sufficiently large $n$. Assume $a = 1/2$ and that x contains $x[n] = u[n]$ for $0 \le n \le 30$. Assuming both $y[-1] = 0$ and $y[-1] = 1/2$, compute the two output signals $y[n]$ for $0 \le n \le 30$. Use `stem` to display both responses. How do they differ?

## 1.6 Continuous-Time Complex Exponential Signals Ⓢ

Before starting this exercise, you are strongly encouraged to work through the Symbolic Math Toolbox tutorial contained in the MATLAB manual. The functions in the Symbolic Math Toolbox can be used to represent, manipulate, and analyze continuous-time signals and systems symbolically rather than numerically. As an example, consider the continuous-time complex exponential signals which have the form $e^{st}$, where $s$ is a complex scalar. Complex exponentials are particularly useful for analyzing signals and systems, since they form the building blocks for a large class of signals. Two familiar signals which can be expressed as a sum of complex exponentials are cosine and sine. For example, by setting $s = \pm i\omega t$, we obtain

$$\cos(\omega t) = \frac{1}{2}\left(e^{i\omega t} + e^{-i\omega t}\right), \tag{1.7}$$

$$\sin(\omega t) = \frac{1}{2i}\left(e^{i\omega t} - e^{-i\omega t}\right). \tag{1.8}$$

In this exercise, you will be asked to use the Symbolic Math Toolbox to represent some basic complex exponential and sinusoidal signals. You will also plot these signals using `ezplot`, the plotting routine of the Symbolic Math Toolbox.

### Basic Problems

(a). Consider the continuous-time sinusoid

$$x(t) = \sin(2\pi t/T).$$

A symbolic expression can be created to represent $x(t)$ within MATLAB by executing

```
>>  x = sym('sin(2*pi*t/T)');
```

The variables of x are the single character strings 't' and 'T'. The function `ezplot` can be used to plot a symbolic expression which has only one variable, so you must set the fundamental period of $x(t)$ to a particular value. If you desire $T = 5$, you can use `subs` as follows:

```
>>  x5 = subs(x,5,'T');
```

Thus x5 is a symbolic expression for $\sin(2\pi t/5)$. Create the symbolic expression for x5 and use `ezplot` to plot two periods of $\sin(2\pi t/5)$, beginning at $t = 0$. If done correctly, your plot should be as shown in Figure ??.

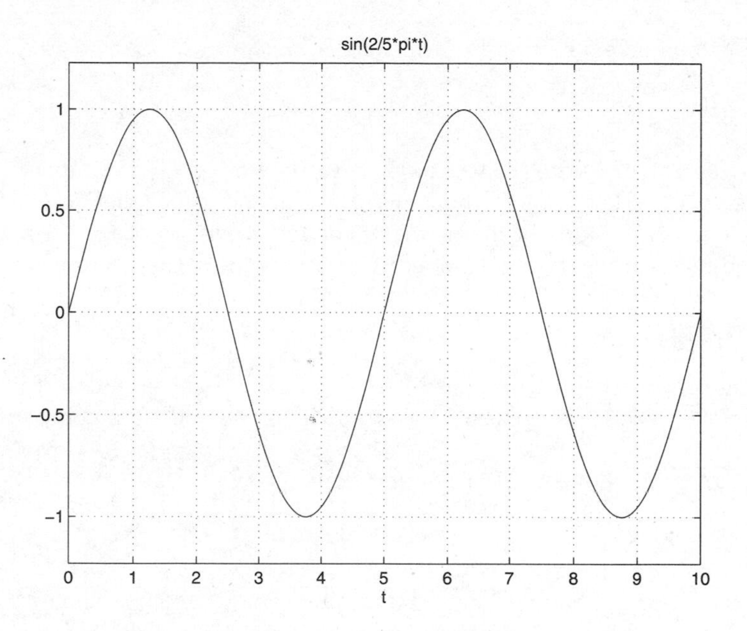


**Figure 1.2.** Two periods of the signal $\sin(2\pi t/5)$.

(b). Create a symbolic expression for the signal

$$x(t) = \cos(2\pi t/T)\,\sin(2\pi t/T)\,.$$

For $T = 4$, 8, and 16, use `ezplot` to plot the signal on the interval $0 \leq t \leq 32$. What is the fundamental period of $x(t)$ in terms of $T$?

## Intermediate Problem

The response of some underdamped systems to an impulsive input can be modeled by

$$e^{-at}\cos(2\pi t/T)\,u(t)\,.$$

An example of a physical system which might generate such a signal is the sound generated by striking a bell. This sound is well approximated by a single tone whose magnitude decays with time. For underdamped systems, the quality[1]

$$Q = \frac{(2\pi/T)}{2a} \tag{1.9}$$

is often used to quantify the resonance of the system. The resonance is a measure of the number of oscillations in the impulse response before the response effectively dies out. For the bell example, the time at which the response dies out could be defined as the time at which the sound becomes inaudible.

[1]The definition of quality in Eq. (??) is actually an approximation of the quality defined in *Signals and Systems* by Oppenheim and Willsky, and is valid only when $Ta \ll \pi$.

(c). Create a symbolic expression for the signal

$$x(t) = e^{-at}\cos(2\pi t)\,.$$

For $a = 1/2$, $1/4$, and $1/8$, use `ezplot` to determine $t_d$, the time at which $|x(t)|$ last crosses 0.1. Define $t_d$ as the time at which the signal dies out. Use `ezplot` to determine for each value of $a$ how many complete periods of the cosine occur before the signal dies out. Does the number of periods appear to be proportional to $Q$?

## Advanced Problems

In the following problems you will write M-files for extracting the real and imaginary components, or the magnitude and phase, of a symbolic expression for a complex signal.

(d). Store in `x` a symbolic expression for the signal

$$x(t) = e^{i2\pi t/16} + e^{i2\pi t/8}\,.$$

The function `ezplot` cannot be used directly for plotting $x(t)$, since $x(t)$ is a complex signal. Instead, the real and imaginary components must be extracted and then plotted separately.

(e). Write a function `xr=sreal(x)` which returns a symbolic expression `xr` representing the real part of $x(t)$. If your function is working properly, `ezplot(xr)` will plot the real component of $x(t)$. Similarly, write a function `xi=simag(x)` which returns a symbolic expression `xi` representing the imaginary component of $x(t)$. The first line of the M-file `sreal.m` should be

```
function xr = sreal(x)
```

You can then use `compose('real(x)',x)` to create a symbolic expression for the real component of $x(t)$. Use `ezplot` and the functions you created to plot the real and imaginary components of $x(t)$ on the interval $0 \le t \le 32$. Use a separate plot for each component. What is the fundamental period of $x(t)$?

(f). For `x` containing the symbolic expression for $x(t)$, create two functions `xm=sabs(x)` and `xa=sangle(x)` which create symbolic expressions representing the magnitude and phase, respectively, of $x(t)$.

(g). Consider again $x(t)$ as defined in Part ??. Use `ezplot` and the functions you created to plot the magnitude and phase of $x(t)$ on the interval $0 \le t \le 32$. Use separate plots for the magnitude and phase. Why is the phase plot discontinuous?

## ■ 1.7 Transformations of the Time Index for Continuous-Time Signals Ⓢ

This exercise will allow you to examine the effect of various transformations of the independent variable of continuous-time signals using MATLAB's Symbolic Math Toolbox.

Specifically, you will look at the effect of these transformations on a ramp-shaped pulse signal

$$f(t) = t\big(u(t) - u(t-2)\big), \tag{1.10}$$

where $u(t)$ is the unit step signal

$$u(t) = \begin{cases} 1, & t \geq 0, \\ 0, & t < 0. \end{cases}$$

The Symbolic Math Toolbox in MATLAB calls the unit step function **Heaviside**. The function `ezplot` can only plot functions which are both in the Symbolic Math and main MATLAB toolboxes. Since **Heaviside** is only in the Symbolic Math Toolbox, you will need to create an M-file called **Heaviside.m** in your working directory. The contents of this file are as follows:

```
function f = Heaviside(t)
% HEAVISIDE Unit Step function
% f = Heaviside(t) returns a vector f the same size as
% the input vector, where each element of f is 1 if the
% corresponding element of t is greater than or equal to
% zero.
f = (t>=0);
```

If you have defined this function properly, you should be able to duplicate the following example:

```
>> Heaviside([-1:0.2:1])
ans =
     0     0     0     0     0     1     1     1     1     1     1
```

## Intermediate Problems

(a). Use **Heaviside** to define **f** to be a symbolic expression for $f(t)$ as specified in Eq. (??). Plot this symbolic expression using `ezplot`.

(b). The expressions below define a set of continuous-time signals in terms of $f(t)$. For each of the following signals, state how you expect it to be related to $f(t)$, e.g., "delayed by 7," "flipped then advanced by 16":

$$\begin{aligned} g_1(t) &= f(-t)\,, \\ g_2(t) &= f(t+1)\,, \\ g_3(t) &= f(t-3)\,, \\ g_4(t) &= f(-t+1)\,, \\ g_5(t) &= f(-2t+1)\,. \end{aligned}$$

(c). Use the Symbolic Math Toolbox function `subs` and the symbolic expression `f` you defined in Part ?? to define symbolic expressions in MATLAB called `g1` through `g5` to represent the signals in Part ??. Plot each signal using `ezplot` and state whether or not the plot agrees with your prediction from Part ??.

## 1.8 Energy and Power for Continuous-Time Signals Ⓢ

For a continuous-time signal $x(t)$, the energy over the interval $-a \leq t \leq a$ is often defined as

$$E_a = \int_{-a}^{a} |x(t)|^2 dt\,, \qquad a \geq 0\,,$$

where $|x|^2 = x\,x^*$ and $x^*$ is the complex conjugate of $x$. Thus, for a periodic signal with fundamental period $T$, $E_{T/2}$ contains the signal energy over one period. The energy in the entire signal is defined as

$$E_\infty = \lim_{a\to\infty} E_a\,,$$

if the limit exists. While most signals in practice have finite energy, many of the continuous-time signals used as conceptual tools for signals and systems do not. For example, any periodic signal has infinite energy. For these signals, a more useful measure is average power, which is simply energy divided by the length of the time interval. Thus, the time-average power over the interval $-a \leq t \leq a$ is

$$P_a = \frac{E_a}{2a}\,, \qquad a > 0\,,$$

and the time-average power of the entire signal is

$$P_\infty = \lim_{a\to\infty} P_a\,,$$

if the limit exists. In this problem, you will consider how $P_\infty$ and $E_{T/2}$ are related for periodic signals.

### Basic Problems

(a). Create symbolic expressions for each of the following three signals:

$$x_1(t) = \cos(\pi t/5)\,,$$
$$x_2(t) = \sin(\pi t/5)\,,$$
$$x_3(t) = e^{i2\pi t/3} + e^{i\pi t}\,.$$

These expressions will have `'t'` as a variable. You might want to use the function `symadd` when creating the symbolic expression for $x_3(t)$, though it is not necessary.

(b). Use `ezplot` to plot two periods of each signal. If the signal is complex, be sure to plot the real and imaginary components separately. The axes of your plots should be appropriately labeled. Hint: You can extract the real component of a symbolic expression using `compose('real(x)',x)`. If you have done Exercise ??, use the functions you created there.

## Intermediate Problems

(c). Define `E1, E2,` and `E3` to be the symbolic expressions containing $E_a$ for the signals $x_1(t)$, $x_2(t)$, and $x_3(t)$, respectively. You should use `int` with the symbolic expressions `'a'` and `'-a'` as the limits of integration. Also, to obtain the complex conjugate of the symbolic expression `x`, you can type `subs(x,'i','-i')`.

(d). Use the symbolic expression for each signal to evaluate $E_{T/2}$, the energy in a single period of the signal. Your answers should be numbers, not expressions. You will probably need to use both `subs` and `numeric`. For each symbolic expression, use `ezplot` to plot $E_a$ as a function of $a$ for $0 \leq a \leq 30$. How does the energy change as the interval length increases? What values do you expect for $E_\infty$?

(e). Define `P1, P2,` and `P3` to be the symbolic expressions containing $P_a$ for the signals $x_1(t)$, $x_2(t)$, and $x_3(t)$, respectively. Create each of these symbolic expressions and use `ezplot` to display $P_a$ for $0.1 \leq a \leq 60$. Remember that $P_a$ is undefined for $a = 0$. How does $P_a$ behave as $a$ increases? Estimate from your plots the value of $P_\infty$ for each signal. For each signal, how does $P_\infty$ compare to $(E_{T/2})/T$? Clearly explain how you could have anticipated this result from the definitions of $P_\infty$ and $E_{T/2}$.

(f). Why would you expect $P_a$ for $x_1(t)$ and $P_a$ for $x_2(t)$ to converge to the same value? Use `symadd` to add the symbolic expressions `P1` and `P2`. Be sure to invoke `simple` to simplify this expression as much as possible. What expression do you obtain? Explain how you could have anticipated this result.

Chapter 2

# Linear Time-Invariant Systems

The exercises in this chapter cover many of the properties of linear time-invariant (LTI) systems. The exercises in Chapter 1 provided an introduction to the basic concepts involved in using MATLAB to represent signals and systems. You developed the necessary tools for dealing with both numerical and symbolic expressions and learned much of the basic syntax of MATLAB functions and variable expressions. This provided some hands-on experience with many of the properties of basic signals and systems. Two of these properties, linearity and time-invariance, will be the central focus of this chapter. In discrete time, linearity provides the ability to completely characterize a system in terms if its response $h_k[n]$ to signals of the form $\delta[n-k]$ for all $k$. If a linear system is also time-invariant, then the responses $h_k[n]$ satisfy $h_k[n] = h[n-k]$. The combination of linearity and time-invariance therefore allows a system to be completely described by its impulse response $h[n]$, since the output of the system $y[n]$ is related to the input $x[n]$ through the convolution sum

$$y[n] = \sum_{m=-\infty}^{\infty} h[n-m]x[m]. \tag{2.1}$$

Similarly, the output $y(t)$ of a continuous-time LTI system is related to the input $x(t)$ and the impulse response $h(t)$ through the convolution integral

$$y(t) = \int_{-\infty}^{\infty} h(t-\tau)x(\tau)d\tau. \tag{2.2}$$

The numerical capabilities of MATLAB can help you to understand some of the basic properties of the convolution operation. The first three tutorials in this chapter explain how to use MATLAB to compute the output of LTI systems using the functions `conv`, `filter`, and `lsim`. Some of the exercises in this chapter will cover numerical issues involved in the approximation of continuous-time systems. Others, like Exercises ?? and ?? explore in greater detail many of the properties of discrete-time LTI systems. In Exercise ??, you will learn how to use MATLAB to implement certain noncausal LTI systems that have finite-length impulse responses. Exercise ?? describes a method for implementing discrete-time convolution when one of the signals is very long. Exercise ?? describes a numerical approximation to continuous-time convolution. In Exercise ??, you will learn how the pulse response of continuous-time systems can be used to characterize their behavior. Finally, in Exercise ?? you will learn how to perform echo cancellation on speech signals using inverse filtering.

## 2.1 Tutorial: `conv`

The MATLAB function `conv` computes the convolution sum

$$y[n] = \sum_{m=-\infty}^{\infty} h[m]x[n-m], \tag{2.3}$$

assuming that $x[n]$ and $h[n]$ are finite-length sequences. If $x[n]$ is nonzero only on the interval $n_x \le n \le n_x + N_x - 1$ and $h[n]$ is nonzero only on the interval $n_h \le n \le n_h + N_h - 1$, then $y[n]$ can be nonzero only on the interval

$$(n_x + n_h) \le n \le (n_x + n_h) + N_x + N_h - 2\,, \tag{2.4}$$

meaning that `conv` need only compute $y[n]$ for the $N_x + N_h - 1$ samples on this interval. If `x` is an $N_x$–dimensional vector containing $x[n]$ on the interval $n_x \le n \le n_x + N_x - 1$ and `h` is an $N_h$–dimensional vector containing $h[n]$ on the interval $n_h \le n \le n_h + N_h - 1$, then `y=conv(h,x)` returns in `y` the $N_x + N_h - 1$ samples of $y[n]$ on the interval in Eq. (??). However, `conv` does not return the indices of the samples of $y[n]$ stored in `y`, which makes sense because the intervals of `x` and `h` are not input to `conv`. Instead, you are responsible for keeping track of these indices, and will be shown how to do this in this tutorial.

(a). Consider the finite-length signal

$$x[n] = \begin{cases} 1\,, & 0 \le n \le 5\,, \\ 0\,, & \text{otherwise}\,. \end{cases} \tag{2.5}$$

Analytically determine $y[n] = x[n] * x[n]$.

(b). Compute the nonzero samples of $y[n] = x[n] * x[n]$ using `conv`, and store these samples in the vector `y`. Your first step should be to define the vector `x` to contain the samples of $x[n]$ on the interval $0 \le n \le 5$. Also construct an index vector `ny`, where `ny(i)` contains the index of the sample of $y[n]$ stored in the $i$-th element of `y`, i.e., `y(i)` $= y[$`ny(i)`$]$. For example, `ny(1)` should contain $n_x + n_x$, where $n_x$ is the first nonzero index of $x[n]$. Plot your results using `stem(ny,y)`, and make sure that your plot agrees with the signal determined in Part ??. As a check, your plot should also agree with Figure ??.

(c). Consider the finite-length signal

$$h[n] = \begin{cases} n\,, & 0 \le n \le 5\,, \\ 0\,, & \text{otherwise}\,. \end{cases} \tag{2.6}$$

Analytically compute $y[n] = x[n] * h[n]$. Next, compute `y` using `conv`, where your first step should be to define the vector `h` to contain $h[n]$ on the interval $0 \le n \le 5$. Again construct a vector `ny` which contains the interval of $n$ for which `y` contains $y[n]$. Plot your results using `stem(ny,y)`. As a check, your plot should agree with Figure ?? and your analytical derivation.

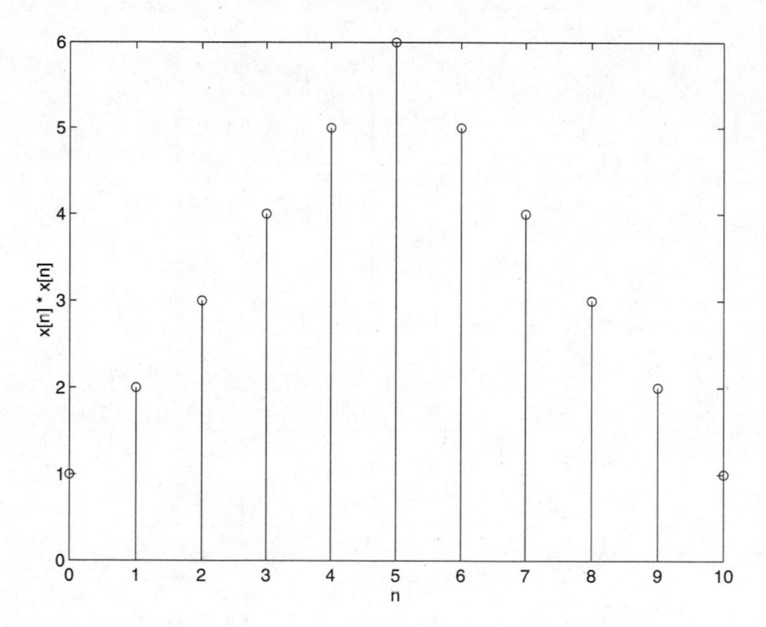


**Figure 2.1.** Plot of the signal $y[n] = x[n] * x[n]$.

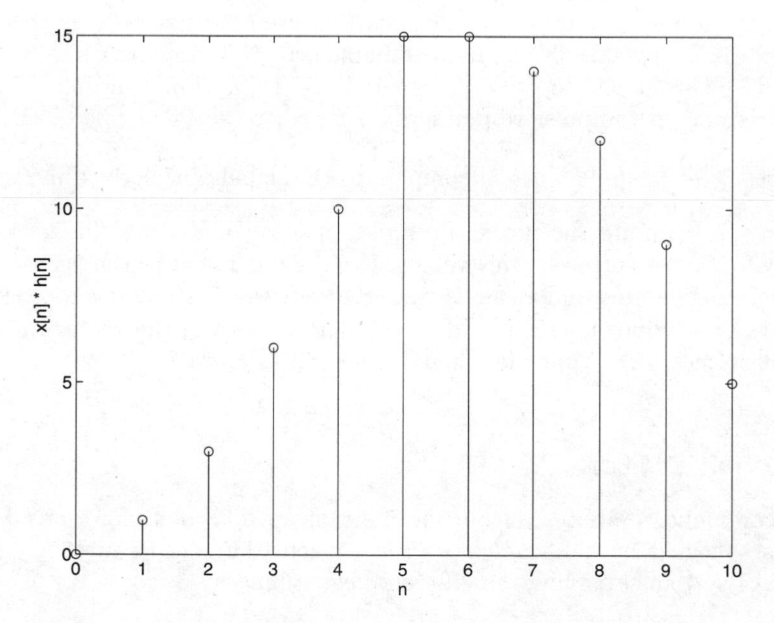


**Figure 2.2.** Plot of the signal $y[n] = x[n] * h[n]$.

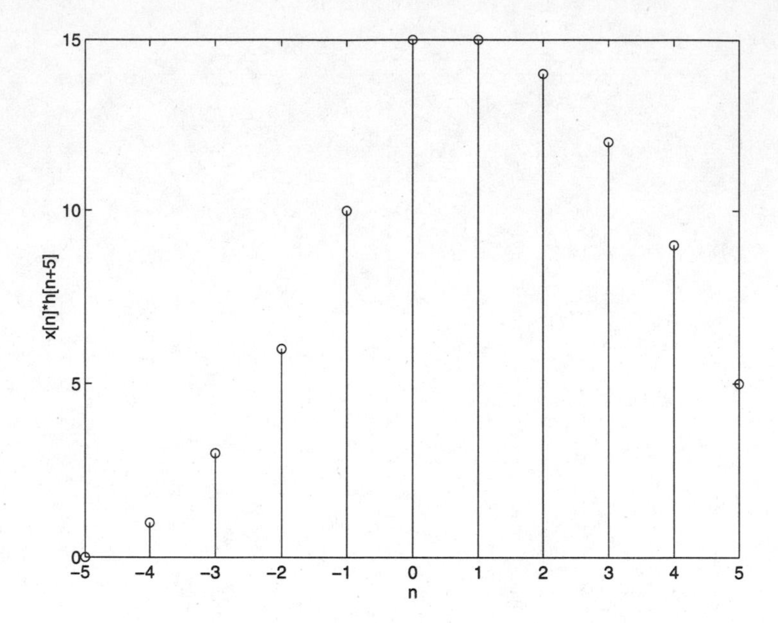


**Figure 2.3.** Plot of the signal $y_2[n] = x[n] * h[n+5]$.

For the example just considered—implementing $y[n] = x[n] * h[n]$ using `conv`—the signal $h[n]$ can be viewed as the impulse response of a linear time-invariant system for which $x[n]$ is the system input and $y[n]$ is the system output. Because $h[n]$ is zero for $n < 0$, this system is causal. However, `conv` can also be used to implement LTI systems which are noncausal. You must still be careful to keep track of the indices of $x[n]$, $h[n]$, and $y[n]$. For example, consider the system with impulse response $h[n+5]$, where $h[n]$ is defined in Eq. (**??**).

(d). How does $y_2[n] = x[n] * h[n+5]$ compare to the signal $y[n]$ derived in Part **??**?

(e). Use `conv` to compute the nonzero samples of $y_2[n]$, and store these samples in the vector `y2`. If done correctly, this vector should be identical to the vector `y` computed in Part **??**. The only difference is that the indices of the values stored in `y2` have changed. Determine this set of indices, and store them in the vector `ny2`. Plot $y_2[n]$ using `stem(ny2,y2)`. Your plot should agree with Figure **??**.

## ■ 2.2 Tutorial: `filter`

The `filter` command computes the output of a causal, LTI system for a given input when the system is specified by a linear constant-coefficient difference equation. Specifically, consider an LTI system satisfying the difference equation

$$\sum_{k=0}^{K} a_k y[n-k] = \sum_{m=0}^{M} b_m x[n-m], \tag{2.7}$$

where $x[n]$ is the system input and $y[n]$ is the system output. If x is a MATLAB vector containing the input $x[n]$ on the interval $n_x \leq n \leq n_x + N_x - 1$ and the vectors a and b contain the coefficients $a_k$ and $b_m$, then `y=filter(b,a,x)` returns the output of the causal LTI system satisfying

$$\sum_{k=0}^{K} \mathtt{a}(k+1)\mathtt{y}(n-k) = \sum_{m=0}^{M} \mathtt{b}(m+1)\mathtt{x}(n-m)\,. \tag{2.8}$$

Note that $\mathtt{a}(k+1) = a_k$ and $\mathtt{b}(m+1) = b_m$, since MATLAB requires that all vector indices begin at one. For example, to specify the system described by the difference equation $y[n] + 2y[n-1] = x[n] - 3x[n-1]$, you would define these vectors as `a=[1 2]` and `b=[1 -3]`.
The output vector y returned by `filter` contains samples of $y[n]$ on the same interval as the samples in x, i.e., $n_x \leq n \leq n_x + N_x - 1$, so that both x and y contain $N_x$ samples. Note, however, that `filter` needs $x[n]$ for $n_x - M \leq n \leq n_x - 1$ and $y[n]$ for $n_x - K \leq n \leq n_x - 1$ in order to compute the first output value $y[n_x]$. The function `filter` assumes that these samples are equal to zero.

(a). Define coefficient vectors `a1` and `b1` to describe the causal LTI system specified by $y[n] = 0.5x[n] + x[n-1] + 2x[n-2]$.

(b). Define coefficient vectors `a2` and `b2` to describe the causal LTI system specified by $y[n] = 0.8y[n-1] + 2x[n]$.

(c). Define coefficient vectors `a3` and `b3` to describe the causal LTI system specified by $y[n] - 0.8y[n-1] = 2x[n-1]$.

(d). For each of these three systems, use `filter` to compute the response $y[n]$ on the interval $1 \leq n \leq 4$ to the input signal $x[n] = n\,u[n]$. You should begin by defining the vector `x=[1 2 3 4]`, which contains $x[n]$ on the interval $1 \leq n \leq 4$. The result of using `filter` for each system is shown below:

```
>> x = [1 2 3 4];
>> y1 = filter(b1,a1,x)
y1 =
    0.5000    2.0000    5.5000    9.0000
>> y2 = filter(b2,a2,x)
y2 =
    2.0000    5.6000   10.4800   16.3840
>> y3 = filter(b3,a3,x)
y3 =
         0    2.0000    5.6000   10.4800
```

From `y1(1)=0.5`, you can see that `filter` has set $x[0]$ and $x[-1]$ equal to zero, since both of these samples are needed to determine $y_1[1]$.

The function filter can also be used to perform discrete-time convolution. Consider the class of systems satisfying Eq. (??) when $a_k = \delta[k]$. In this case, Eq. (??) becomes

$$y[n] = \sum_{m=0}^{M} b_m x[n-m]\,. \tag{2.9}$$

If we define the following finite-length signal

$$b[m] = \begin{cases} b_m\,, & 0 \le m \le M\,, \\ 0\,, & \text{otherwise}\,, \end{cases}$$

then Eq. (??) can be rewritten as

$$y[n] = \sum_{m=0}^{M} b[m]x[n-m] = \sum_{m=-\infty}^{\infty} b[m]x[n-m]\,. \tag{2.10}$$

Note the similarity between Eq. (??) and Eq. (??)—the filter given by Eq. (??) is a convolution. The signal $b[m]$ is the impulse response of the LTI system which satisfies Eq. (??). Because $b[m]$ has finite length, such systems are called finite-length impulse response (FIR) filters.
To illustrate how to use filter to implement a discrete-time convolution, consider the convolution of $x[n]$ in Eq. (??) with $h[n]$ in Eq. (??).

(e). Store $x[n]$ and $h[n]$ on the interval $0 \le n \le 5$ in the vectors x and h.

(f). To use filter, the impulse response $h[n]$ must be mapped to the coefficients of the difference equation in Eq. (??), i.e., $b_m = h[m]$ for $0 \le n \le 5$. In other words, the difference equation coefficients are given by b=h and a=1. Use y=filter(h,1,x) to compute the output of this difference equation on the interval $0 \le n \le 5$, and set ny=[0:5]. Remember that filter returns a vector y with the same number of samples as x. Plot your results using stem(ny,y). Your plot should agree with Figure ??.

If filter is to return the same result as conv(h,x), then the input to filter must contain 11 samples of $x[n]$. (Remember that conv returns a vector of length $N_x + N_h - 1$, where $N_x$ is the length of x and $N_h$ is the length of h.)

(g). Define a vector x2 to contain $x[n]$ on the interval $0 \le n \le 10$, and use

```
>> y2=filter(h,1,x2);
```

to compute the convolution on this interval. Plot your results using stem([0:10],y2), and verify that your plot agrees with Figure ??.

Like conv, filter can also be used to implement an LTI system which has a noncausal impulse response. Again, it is important to keep track of the indices of the input, impulse

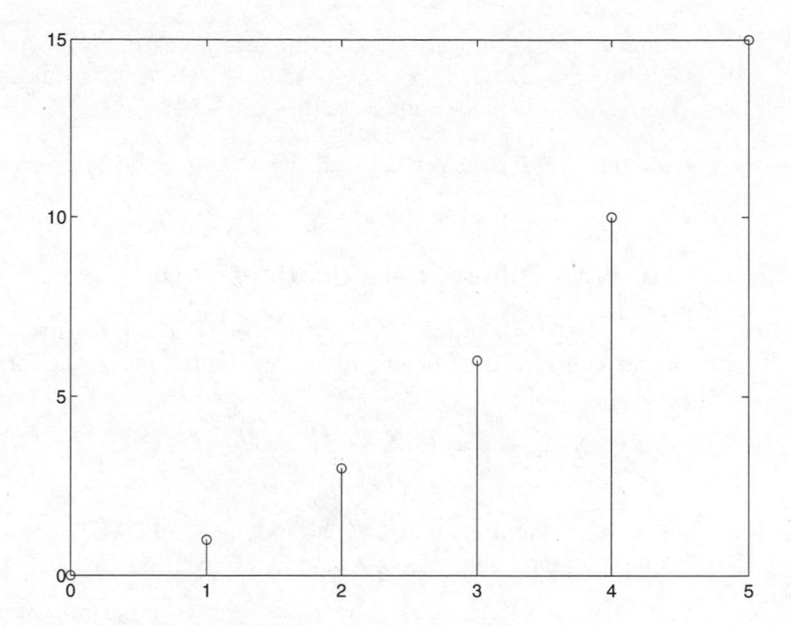


**Figure 2.4.** Plot of the signal $y[n] = x[n] * h[n]$ over the interval $0 \leq n \leq 5$.

response, and output. To illustrate how, assume that $h[n]$ in Eq. (??) is replaced by $h[n+L]$ for some integer $L$. The convolution sum becomes

$$y_2[n] = \sum_{m=-\infty}^{\infty} h[m+L]x[n-m],$$

which, upon substitution of $m' = m + L$, gives

$$\begin{aligned} y_2[n] &= \sum_{m'=-\infty}^{\infty} h[m']x[n-(m'-L)] \\ &= \sum_{m'=-\infty}^{\infty} h[m']x[(n+L)-m'] \\ &= y[n+L]. \end{aligned}$$

In other words, an advance in the impulse response by $L$ samples merely advances the output by $L$ samples. For $L < 0$, this corresponds to a delay. Therefore, if `x` contains $x[n]$ on the interval $0 \leq n \leq N_x - 1$ and `h` contains $h[n]$ on the interval $-L \leq n \leq N_h - 1 - L$, then `y=filter(h,1,x)` will return $y[n]$ on the interval $-L \leq n \leq N_x - 1 - L$. Note that `y` still has the same number of samples as `x`, only the samples represented by `y` have advanced by $L$ samples.

(h). Consider the impulse response $h_2[n] = h[n+5]$, where $h[n]$ is defined in Eq. (??). Store $h_2[n]$ on the interval $-5 \leq n \leq 0$ in the vector `h2`.

(i). Execute the command `y2=filter(h2,1,x)` and create a vector `ny2` which contains indices of the samples of $y_2[n] = h_2[n] * x[n]$ stored in `y2`. Plot your result using `stem(ny2,y2)`. How does this plot compare with Figure ???

(j). Create a vector `x2` such that `filter(h2,1,x2)` returns all the nonzero samples of $y_2[n]$.

## 2.3 Tutorial: `lsim` with Differential Equations

The function `lsim` can be used to simulate the output of continuous-time, causal LTI systems described by linear constant-coefficient differential equations of the form

$$\sum_{k=0}^{N} a_k \frac{d^k y(t)}{dt^k} = \sum_{m=0}^{M} b_m \frac{d^m x(t)}{dt^m}. \tag{2.11}$$

To use `lsim`, the coefficients $a_k$ and $b_m$ must be stored in MATLAB vectors `a` and `b`, respectively, in descending[1] order of the indices $k$ and $m$. Rewriting Eq. (??) in terms of the vectors `a` and `b` gives

$$\sum_{k=0}^{N} \mathtt{a}(N+1-k) \frac{d^k y(t)}{dt^k} = \sum_{m=0}^{M} \mathtt{b}(M+1-m) \frac{d^m x(t)}{dt^m}. \tag{2.12}$$

Note that `a` must contain $N+1$ elements, which might require appending zeros to `a` to account for coefficients $a_k$ that equal zero. Similarly, the vector `b` must contain $M+1$ elements. With `a` and `b` defined as in Eq. (??), executing

```
>> y = lsim(b,a,x,t);
```

simulates the response of Eq. (??) to the input signal specified by the vectors `x` and `t`. The vector `t` contains the time samples for the input and output, `x` contains the values of the input $x(t)$ at each time in `t`, and `y` contains the simulated values of the output $y(t)$ at each time in `t`. The accuracy of the simulated values depends upon how well `x` and `t` represent the true function $x(t)$.

While this tutorial does not describe the numerical methods used by `lsim` to compute `y`, it is important to know how `lsim` interprets the inputs `x` and `t`. Basically, `lsim` interpolates the pair `t,x` in much the same way as does `plot`. For instance, consider the plot produced by the following code:

```
>> t = [0 1 2 5 8 9 10];
>> x = [0 0 0 3 0 0 0];
>> plot(t,x)
```

The plot is given in Figure ??. The function `lsim(b,a,x,t)` will consider $x(t)$ to be

[1] As noted in Tutorial ??, the vectors `a` and `b` for the function `filter` contain $a_k$ and $b_m$ in ascending order of the indices $k$ and $m$, rather than descending order as used by `lsim`.

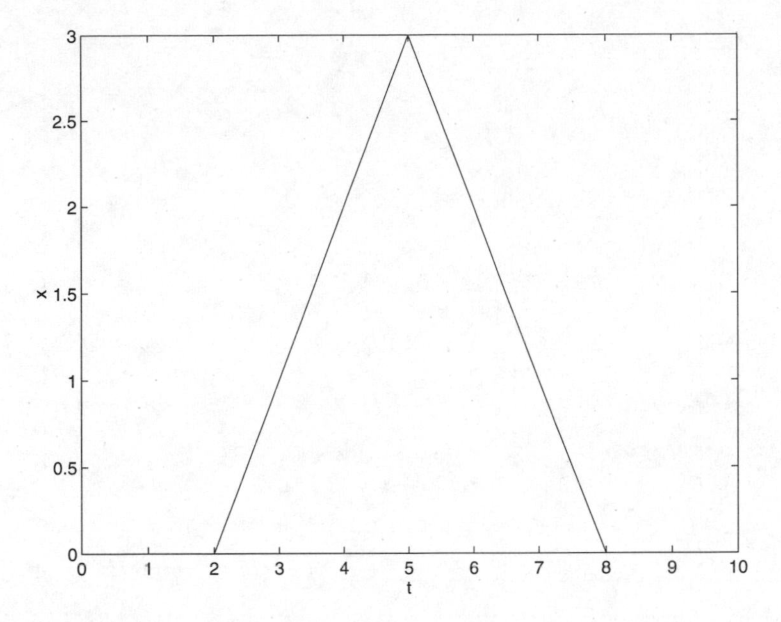


**Figure 2.5.** The linear interpolation used by `plot(t,x)` is similar to that used by `lsim` to create its input function.

equal to

$$x(t) = \begin{cases} 3 - |t-5|\,, & 2 \le t \le 8\,, \\ 0\,, & \text{otherwise}\,. \end{cases}$$

on the interval $0 \le t \le 10$. Thus the linear interpolation of the points specified by the pair `[t(n),x(n)]` is the continuous-time function $x(t)$ which `lsim` uses as the input to Eq. (??). Consider the causal LTI system described by the first-order differential equation

$$\frac{dy(t)}{dt} = -\frac{1}{2}\,y(t) + x(t)\,. \tag{2.13}$$

The step response of this system can be computed by first defining the input step function as

```
>> t = [0:10];
>> x = ones(1,length(t));
```

The simulated step response can then be computed and plotted by executing

```
>> b = 1;
>> a = [1 0.5];
>> s = lsim(b,a,x,t);
>> plot(t,s,'y--')
```

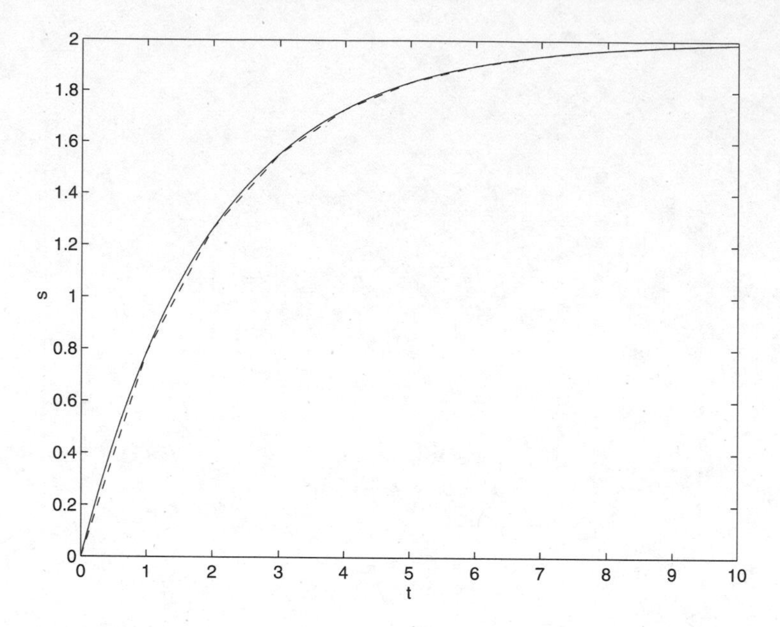


**Figure 2.6.** For the causal LTI system described by Eq. (**??**), the dotted line is a plot of the step response computed by `lsim`. The solid line is the true step response.

The plot is shown in Figure **??**, where the solid line represents the actual step response

$$s(t) = 2\left(1 - e^{-t/2}\right) u(t)\,. \tag{2.14}$$

Note that at each value of `t`, the step response computed by `lsim` is essentially identical to the true step response. The only difference is in the interpolation produced by `plot`. The function `lsim` will return more samples of $s(t)$ if the samples in `t` are chosen more closely spaced, e.g., `t=[0:0.1:10]`.

(a). On your own, use `lsim` to compute the response of the causal LTI system described by

$$\frac{dy(t)}{dt} = -2\,y(t) + x(t)\,. \tag{2.15}$$

to the input $x(t) = u(t-2)$. Your response should look like the plot in Figure **??**, which is computed using `t=[0:0.5:10]`.

While `lsim` can simulate the response of Eq. (**??**) to any input which can be approximated by linear interpolation, the functions `impulse` and `step` can be used to compute the impulse and step responses of such systems. With the vectors `t`, `b`, and `a` defined by

```
>> t = [0:1:10];
>> b = 1;
>> a = [1 0.5];
```

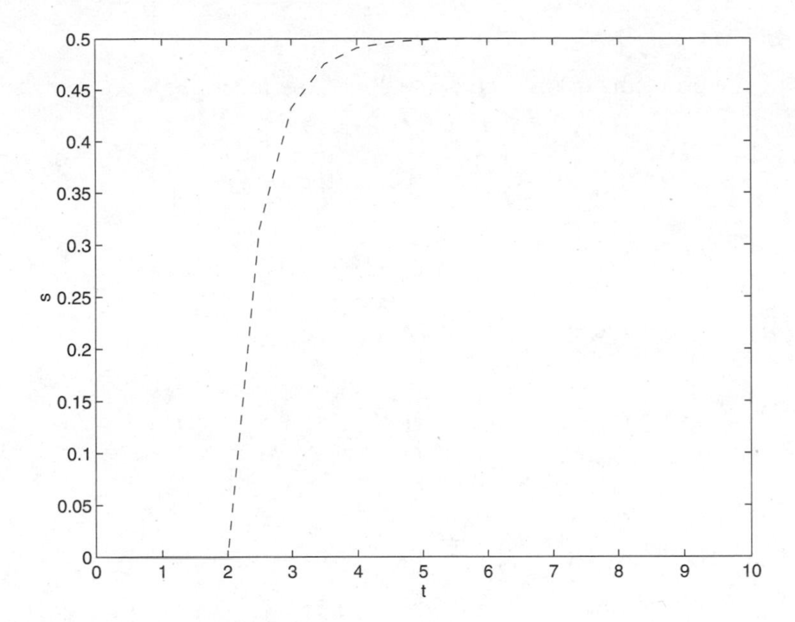


**Figure 2.7.** The response of Eq. (??) to $u(t-2)$, as computed by `lsim` when the time vector is `t=[0:0.5:10]`.

typing

```
>> s = step(b,a,t);
>> h = impulse(b,a,t);
```

will return the step and impulse responses in the vectors `s` and `h`, respectively. Note that the inputs `a` and `b` to `step` and `impulse` have the same form required by `lsim`.

(b). Use `step` and `impulse` to compute the step and impulse responses of the causal LTI system characterized by Eq. (??). Compare the step response computed by `step` with that shown in Figure ??. Compare the signal returned by `impulse` with the exact impulse response, given by the derivative of $s(t)$ in Eq. (??).

## ■ 2.4 Properties of Discrete-Time LTI Systems

In this exercise, you will verify the commutative, associative and distributive properties of convolution for a specific set of signals. In addition, you will examine the implications of these properties for series and parallel connections of LTI systems. The problems in this exercise will assume that you are comfortable and familiar with the `conv` function described in Tutorial ??. Although the problems in this exercise solely explore discrete-time systems, the same properties are also valid for continuous-time systems.

## Basic Problems

(a). Many of the problems in this exercise will use the following three signals:

$$x_1[n] = \begin{cases} 1, & 0 \le n \le 4, \\ 0, & \text{otherwise}, \end{cases}$$

$$h_1[n] = \begin{cases} 1, & n = 0, \\ -1, & n = 1, \\ 3, & n = 2, \\ 1, & n = 4, \\ 0, & \text{otherwise}, \end{cases}$$

$$h_2[n] = \begin{cases} 2, & n = 1, \\ 5, & n = 2, \\ 4, & n = 3, \\ -1, & n = 4, \\ 0, & \text{otherwise}. \end{cases}$$

Define the MATLAB vector `x1` to represent $x_1[n]$ on the interval $0 \le n \le 9$, and the vectors `h1` and `h2` to represent $h_1[n]$ and $h_2[n]$ for $0 \le n \le 4$. Also define `nx1` and `nh1` to be appropriate index vectors for these signals. Make appropriately labeled plots of all the signals using `stem`.

(b). The commutative property states that the result of a convolution is the same regardless of the order of the operands. This implies that the output of an LTI system with impulse response $h[n]$ and input $x[n]$ will be the same as the output of an LTI system with impulse response $x[n]$ and input $h[n]$. Use `conv` with `h1` and `x1` to verify this property. Is the output of `conv` the same regardless of the order of the input arguments?

(c). Convolution is also distributive. This means that

$$x[n] * (h_1[n] + h_2[n]) = x[n] * h_1[n] + x[n] * h_2[n] .$$

This implies that the output of two LTI systems connected in parallel is the same as one system whose impulse response is the sum of the impulse responses of the parallel systems. Figure ?? illustrates this property.

Verify the distributive property using `x1`, `h1` and `h2`. Compute the sum of the outputs of LTI systems with impulse responses $h_1[n]$ and $h_2[n]$ when $x_1[n]$ is the input. Compare this with the output of the LTI system whose impulse response is $h_1[n] + h_2[n]$ when the input is $x_1[n]$. Do these two methods of computing the output give the same result?

(d). Convolution also possesses the associative property, i.e.,

$$(x[n] * h_1[n]) * h_2[n] = x[n] * (h_1[n] * h_2[n]) .$$

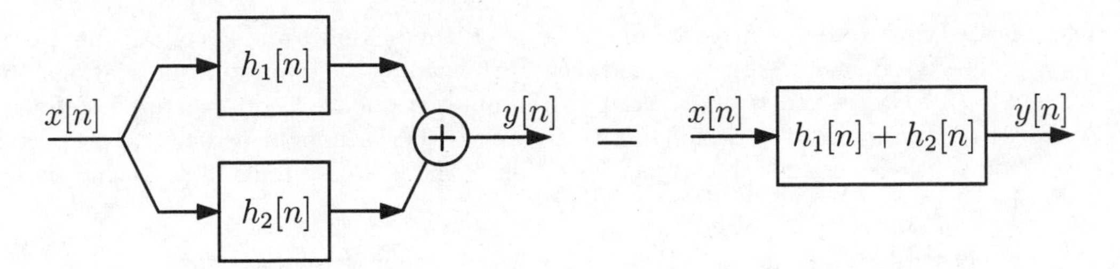


**Figure 2.8.** Distributive property of convolution.

This property implies that the result of processing a signal with a series of LTI systems is equivalent to processing the signal with a single LTI system whose impulse response is the convolution of all the individual impulse responses of the connected systems. Figure **??** illustrates this property for the case of two LTI systems connected in series.

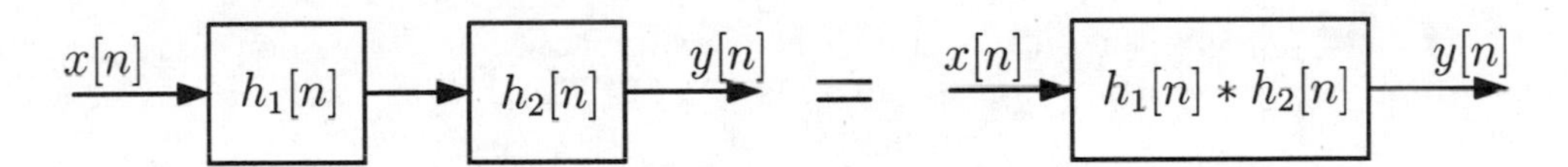


**Figure 2.9.** Associative property of convolution.

Use the following steps to verify the associative property using `x1`, `h1` and `h2`:

- Let $w[n]$ be the output of the LTI system with impulse response $h_1[n]$ shown in Figure **??**. Compute $w[n]$ by convolving $x_1[n]$ and $h_1[n]$.
- Compute the output $y_{d1}[n]$ of the whole system by convolving $w[n]$ with $h_2[n]$.
- Find the impulse response $h_{\text{series}}[n] = h_1[n] * h_2[n]$.
- Convolve $x_1[n]$ with $h_{\text{series}}[n]$ to get the output $y_{d2}[n]$.

Compare $y_{d1}[n]$ and $y_{d2}[n]$. Did you get the same results when you process $x_1[n]$ with the individual impulse responses as when you process it with $h_{\text{series}}[n]$?

## Intermediate Problems

(e). Suppose two LTI systems have impulse responses $h_{e1} = h_1[n]$ and $h_{e2}[n] = h_1[n-n_0]$, where $h_1[n]$ is the same signal defined in Part **??** and $n_0$ is an integer. Let $y_{e1}[n]$ and $y_{e2}[n]$ be the outputs of these systems when $x[n]$ is the input. Use the commutative property to argue that the outputs will be the same if you interchange the input and impulse response of each system. Notice that once you have done this the two systems have the same impulse response and the inputs are delayed versions of the same signal. Based on this observation and time-invariance, argue that $y_{e2}[n] = y_{e1}[n - n_0]$. Use MATLAB to confirm your answer for the case when $n_0 = 2$ and the input $x[n]$ is the signal $x_1[n]$ defined in Part **??**.

(f). Consider two systems connected in series; call them System 1 and System 2. Suppose System 1 is a memoryless system and is characterized by the input/output relationship $y[n] = (n+1)x[n]$, and System 2 is LTI with impulse response $h_{f2}[n] = h_1[n]$ as defined in Part ??. Suppose you decide to investigate whether or not the associative property of convolution holds for the series connection of these two systems by following these steps:

- Let $w[n]$ be the output of System 1 when the input is $x_1[n]$ as defined above. Use `nx1` and `x1` with the termwise multiplication operator `.*` to define a MATLAB vector `w` to represent $w[n]$.
- Use $w[n]$ as the input to System 2, and let the output of that system be $y_{f1}[n]$. Compute `yf1` in MATLAB using `w` and `h1`.
- Let $h_{f1}[n]$ be the output of System 1 when the input to the system is $\delta[n]$. Define a vector `hf1` to represent this signal over the interval $0 \le n \le 4$.
- Let $h_{\text{series}}[n] = h_{f1}[n] * h_{f2}[n]$. Compute a vector `hseries` to represent this signal.
- Let $y_{f2}[n]$ be the output of an LTI system whose impulse response is $h_{\text{series}}[n]$ when the input is $x_1[n]$. Compute `yf2` in MATLAB using `hseries` and `x1`.

Does $y_{f1}[n] = y_{f2}[n]$? If so, why would you expect it to? If not, this means the result of processing a signal with the series connection of Systems 1 and 2 is not equal to the result of processing the signal with a single system whose impulse response is the convolution of the impulse response of System 1 with the impulse response of System 2. Does this violate the associative property of convolution as discussed in Part ???

(g). Consider the parallel connection of two systems; call them System 1 and System 2. System 1 is a memoryless system characterized by the input/output relationship $y[n] = x^2[n]$. System 2 is an LTI system with impulse response $h_{g2}[n] = h_2[n]$ as defined in Part ??. Suppose you were to use the steps below to investigate whether or not the distributive property of convolution held for the parallel connection of these systems:

- Let $y_{ga}[n]$ be the output of System 1 when the input is the signal $x_g[n] = 2\delta[n]$. Define `xg` to represent this input over the interval $0 \le n \le 4$, and use `xg` and the termwise exponentiation operator `.^` to define a MATLAB vector `yga` to represent $y_{ga}[n]$.
- Let $y_{gb}[n]$ be the output of System 2 when $x_g[n]$ is the input, and define `ygb` to represent this signal.
- Let $y_{g1}[n]$ be the sum of $y_{ga}[n]$ and $y_{gb}[n]$, the outputs of the parallel branches. Define the vector `yg1` to represent $y_{g1}[n]$. Note that because `yga` is shorter in length than `ygb`, you will have to extend `yga` with some zeros before you can add the vectors.
- Let $h_{g1}[n]$ be the output of System 1 when the input is $\delta[n]$. Define `hg1` to represent this signal on the interval $0 \le n \le 4$.
- Let $h_{\text{parallel}}[n]$ be $h_{g1}[n] + h_{g2}[n]$. Define `hparallel` to represent this signal.

- Let $y_{g2}[n]$ be the output of the LTI system with impulse response $h_{\text{parallel}}[n]$ when the input is $x_g[n]$. Define a vector yg2 to represent this signal.

Are yg1 and yg2 equal? If so, why would you expect this? If not, has the distributive property of convolution been violated ?

## 2.5 Linearity and Time-Invariance

In this exercise you will become more familiar with the system properties of linearity and time-invariance. In particular, you will be presented with a number of systems, and then be asked to determine if they are linear or time-invariant. This exercise also explores an important property of discrete-time LTI systems: if $h[n]$ is the response to the unit impulse $\delta[n]$, then the response of the system $y[n]$ to any input $x[n]$ is determined by the convolution sum, Eq. (??). An analogous property holds for continuous-time LTI systems.
The problems in this exercise assume that you are familiar with the functions `conv` and `filter`. These functions are explained in Tutorial ?? and Tutorial ??.

### Basic Problems

Consider the systems

$$\begin{aligned} &\text{System 1:} && w[n] = x[n] - x[n-1] - x[n-2]\,, \\ &\text{System 2:} && y[n] = \cos(x[n])\,, \\ &\text{System 3:} && z[n] = n\,x[n]\,, \end{aligned}$$

where $x[n]$ is the input to each system, and $w[n]$, $y[n]$, and $z[n]$ are the corresponding outputs.

(a). Consider the three inputs signals $x_1[n] = \delta[n]$, $x_2[n] = \delta[n-1]$, and $x_3[n] = (\delta[n] + 2\,\delta[n-1])$. For System 1, store in w1, w2, and w3 the responses to the three inputs. The vectors w1, w2, and w3 need to contain the values of $w[n]$ only on the interval $0 \leq n \leq 5$. Use `subplot` and `stem` to plot the four functions represented by w1, w2, w3, and w1+2*w2 within a single figure. Make analogous plots for Systems 2 and 3.

(b). State whether or not each system is linear. If it is linear, justify your answer. If it is not linear, use the signals plotted in Part ?? to supply a counter-example.

(c). State whether or not each system is time-invariant. If it is time-invariant, justify your answer. If it is not time-invariant, use the signals plotted in Part ?? to supply a counter-example.

### Intermediate Problems

In these problems, you will be asked to consider how the impulse response can be used to calculate the step response of an LTI system. Consider the two causal systems defined by the following linear difference equations:

$$\begin{aligned} &\text{System 1:} && y_1[n] = (3/5)\,y_1[n-1] + x[n]\,, \\ &\text{System 2:} && y_2[n] = (3/5)^n\,y_2[n-1] + x[n]\,. \end{aligned}$$

Each system satisfies initial rest conditions, which state that if $x[n] = 0$ for $n \leq n_0$, then $y[n] = 0$ for $n \leq n_0$. Define $h_1[n]$ and $h_2[n]$ to be the responses of Systems 1 and 2, respectively, to the signal $\delta[n]$.

(d). Calculate $h_1[n]$ and $h_2[n]$ on the interval $0 \leq n \leq 19$, and store these responses in `h1` and `h2`. Plot each response using `stem`. Hint: The `filter` function can be used to calculate `h1`. However, System 2 is described by a difference equation with non-constant coefficients; therefore, you must either determine `h2` analytically or use a `for` loop rather than `filter` to calculate `h2`.

(e). For each system, calculate the unit step response on the interval $0 \leq n \leq 19$, and store the responses in `s1` and `s2`. Again, `filter` can be used only to calculate the step response of System 1. Use a `for` loop to calculate `s2`.

(f). Note that $h_1[n]$ and $h_2[n]$ are zero for $n \geq 20$ for all practical purposes. Thus `h1` and `h2` contain all we need to know about the response of each system to the unit impulse. Define $z_1[n] = h_1[n] * u[n]$ and $z_2[n] = h_2[n] * u[n]$, where $u[n]$ is the unit step function. Use `conv` to calculate $z_1[n]$ and $z_2[n]$ on the interval $0 \leq n \leq 19$, and store these calculations in the vectors `z1` and `z2`. You must first define a vector containing $u[n]$ over an appropriate interval, and then select the subset of the samples produced by `conv(h1,u)` and `conv(h2,u)` that represent the interval $0 \leq n \leq 19$. Since you have truncated two infinite-length signals, only a portion of the outputs of `conv` will contain valid sequence values. This issue was also discussed in Exercise ?? Part ??.

(g). Plot `s1` and `z1` on the same set of axes. If the two signals are identical, explain why you could have anticipated this similarity. Otherwise, explain any differences between the two signals. On a different set of axes, plot `s2` and `z2`. Again, explain how you might have anticipated any differences or similarities between these two signals.

## ■ 2.6 Noncausal Finite Impulse Response Filters

In this exercise, you will learn how to implement a class of noncausal LTI systems whose impulse responses have a finite number of nonzero samples. MATLAB assumes all signals start with index one, but by defining a separate vector of time indices for each signal as shown in Tutorial ??, you can implement noncausal systems. The inputs and outputs of these LTI systems are related by the difference equation

$$y[n] = \sum_{m=N_1}^{N_2} b[m]x[n-m]. \tag{2.16}$$

### Basic Problems

(a). Find the impulse response of the LTI system whose input and output satisfy Eq. (??). What can you say about the value of $N_1$ if the system is not causal?

(b). Suppose an LTI system whose impulse response $h[n]$ is only nonzero in the range $N_1 \leq n \leq N_2$ is convolved with a finite-length signal $x[n]$ which is only nonzero on the interval $N_3 \leq n \leq N_4$. The output of the system $y[n] = x[n] * h[n]$ will also be finite-length, with its nonzero samples in the range $N_5 \leq n \leq N_6$. Find expressions for $N_5$ and $N_6$ in terms of $N_1$ through $N_4$.

(c). Let $x[n]$ be the finite-length signal

$$x[n] = \begin{cases} 1, & n = 0, \\ 5, & n = 1, \\ 2, & n = 2, \\ 4, & n = 3, \\ -2, & n = 4, \\ 2, & n = 5, \end{cases}$$

and let $h[n]$ be the impulse response of a noncausal system given by

$$h[n] = \begin{cases} 1 - (|n|/3), & |n| \leq 3, \\ 0, & \text{otherwise}. \end{cases}$$

Define MATLAB vectors `x` and `h` to represent these signals, along with accompanying index vectors `nx` and `nh`. Make appropriately labeled plots of both signals with `stem`.

(d). Compute the output of the LTI system $y[n] = x[n] * h[n]$ using `conv` and the vectors you defined in the previous part. Define the vector `y` to represent this output vector, and define an index vector `ny`. You may find the expressions you derived in Part ?? useful in determining the indices in `ny`. Make an appropriately labeled plot of $y[n]$ using `stem`.

## Intermediate Problems

One possible use for noncausal LTI systems with finite-length impulse responses is to interpolate between the samples of a discrete-time signal. In Chapter ??, you will consider this interpolation problem in more detail. In that context, you will see that there are a number of criteria for which the LTI system with the impulse response $h[n]$ defined in Part ?? may not be the ideal interpolation system. However, in some situations $h[n]$ provides an acceptable form of interpolation.

(e). Consider a new input signal

$$x_u[n] = \begin{cases} x[n/3], & n = 3k \text{ and } k \text{ integer}, \\ 0, & \text{otherwise}, \end{cases}$$

where $x[n]$ is the signal defined above in Part ??.

The process of inserting zeros between the samples of a signal like this is commonly referred to as expansion. Chapter ?? examines this operation in more detail, but for now you do not need to concern yourself with understanding in-depth what expansion does to the signal. Define MATLAB vectors `xu` and `nxu` to represent the expanded signal and its time indices.

(f). Write the difference equation for the system whose impulse response is $h[n]$ from Part ??. Assuming the input to this system is expanded like $x_u[n]$, what can you say about the output of the system when $n = 3k$? How is the output related to $x_u[n]$ when $n \neq 3k$ ?

(g). Let $y_u[n] = x_u[n] * h[n]$. Compute MATLAB vectors to represent $y_u[n]$ and its time indices. Again, you may find the expressions in Part ?? useful in defining the index vector. Make appropriately labeled plots of $x_u[n]$ and $y_u[n]$. Do these plots confirm your answer from Part ???

## 2.7 Discrete-Time Convolution

The convolution of discrete-time sequences $h[n]$ and $x[n]$ is represented mathematically by the expression given in Eq. (??), which can be viewed pictorially as the operation of flipping the time axis of the sequence $h[m]$ and shifting it by $n$ samples, then multiplying $h[n-m]$ by $x[m]$ and summing the resulting product sequence over $m$. This picture arises from the properties of linearity and time-invariance for discrete-time systems. The signal $x[n]$ can be constructed from a linear superposition of delayed and scaled impulses. Since an LTI system can be represented by its response to a single impulse, the output of an LTI system corresponds to the superposition of the responses of the system to each of the delayed and scaled impulses used to construct $x[n]$. Mathematically this results in the convolution sum. In this project you will become more experienced with the use of the MATLAB function `conv` to compute the convolution of discrete-time sequences as described in Tutorial ??.

### Basic Problems

In these problems, you will define some discrete-time signals and the impulse responses of some discrete-time LTI systems. Then the output of the LTI systems can be computed using `conv`.

(a). Since the MATLAB function `conv` does not keep track of the time indices of the sequences that are convolved, you will have to do some extra bookkeeping in order to determine the proper indices for the result of the `conv` function. For the sequences $h[n] = 2\delta[n+1] - 2\delta[n-1]$, and $x[n] = \delta[n] + \delta[n-2]$ construct vectors `h` and `x`. Define $y[n] = x[n] * h[n]$ and compute `y=conv(h,x)`. Determine the proper time indexing for `y` and store this set of time indices in the vector `ny`. Plot $y[n]$ as a function of $n$ using `stem(ny,y)`.

(b). Consider two finite-length sequences $h[n]$ and $x[n]$ that are represented in MATLAB using the vectors `h` and `x`, with corresponding time indices given by `nh=[a:b]` and `nx=[c:d]`. The function call `y=conv(h,x)` will return in the vector `y` the proper sequence values of $y[n] = h[n] * x[n]$; however, you must determine a corresponding set of time indices `ny`. To help you construct `ny`, consider the sequence $h[n] = \delta[n-a] + \delta[n-b]$ and $x[n] = \delta[n-c] + \delta[n-d]$. Determine analytically the convolution $y[n] = h[n] * x[n]$. From your answer, determine what `ny` should be in terms of $a, b, c$, and $d$. To check your result, verify that $y[n]$ is of length $M + N - 1$ when $a = 0$, $b = N - 1$, $c = 0$, and $d = M - 1$.

(c). Consider an input $x[n]$ and unit impulse response $h[n]$ given by

$$x[n] = \left(\frac{1}{2}\right)^{n-2} u[n-2],$$

$$h[n] = u[n+2].$$

If you wish to compute $y[n] = h[n] * x[n]$ using `conv`, you must deal appropriately with the infinite length of both $x[n]$ and $h[n]$. Store in the vector `x` the values of $x[n]$ for $0 \leq n \leq 24$ and store in the vector `h` the values of $h[n]$ for $-2 \leq n \leq 14$. Now store in the vector `y` the result of the function call `conv(h,x)`. Since you have truncated both $h[n]$ and $x[n]$, argue that only a portion of the output of `conv` will be valid. Specify which values in the output are valid and which are not. Determine the values of the parameters $a$, $b$, $c$, and $d$ such that `nx=[a:b]` and `nh=[c:d]`, and use your answer from Part ?? to construct the proper time indices for `y`. Plot $y[n]$ using `stem`, and indicate which values of $y[n]$ are correct, and which values are invalid. Be sure to properly label the time axis for $y[n]$.

## Intermediate Problems

For these problems, we will consider a technique known as block convolution, which is often used in real-time implementations of digital filters such as speech- or music-processing systems where short processing delays are desired. This technique is most useful when processing a very long input sequence with a relatively short filter. The input sequence is broken into short blocks, each of which can be processed independently with relatively little delay. The linearity of convolution guarantees that the superposition of the outputs from all of the individual blocks will equal the convolution of the entire sequence with the impulse response of the filter. The existence of computationally efficient hardware and algorithms for performing the convolution of finite-length sequences makes this technique especially powerful. For this project, you will perform each of the smaller convolutions using the function `conv`.

To illustrate the procedure, assume that you have a filter with a finite-length impulse response $h[n]$ which is nonzero only on the interval $0 \leq n \leq P-1$. Also assume that the input sequence $x[n]$ is 0 for $n < 0$ and that the length of $x[n]$ is significantly greater than $P$. You can break the signal $x[n]$ into segments of length $L$,

$$x[n] = \sum_{r=0}^{\infty} x_r[n-rL],$$

where $L > P$, and

$$x_r[n] = \begin{cases} x[n+rL], & 0 \leq n \leq L-1, \\ 0, & \text{otherwise}, \end{cases}$$

as shown in Figure ??.

(d). For $h[n] = (0.9)^n(u[n] - u[n-10])$ and $x[n] = \cos(n^2)\sin(2\pi n/5)u[n]$, compute $y[n] = h[n] * x[n]$ for $0 \leq n \leq 99$ directly using `conv`. Make a plot of $y[n]$ over this range using `stem`.

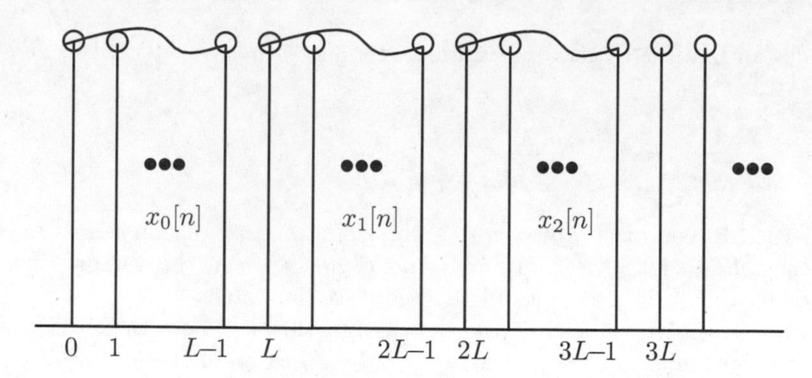


**Figure 2.10.** Block decomposition of $x[n]$.

(e). For $L = 50$, break the sequence $x[n]$ into two sequences, each of length 50. Compute $y_0[n] = h[n] * x_0[n]$, and $y_1[n] = h[n] * x_1[n]$, where $x_0[n]$ contains the first 50 samples of $x[n]$ and $x_1[n]$ contains the second 50 samples of $x[n]$. The form of the output $y[n]$ is given by

$$y[n] = x[n] * h[n] = y_0[n] + y_1[n-k].$$

Determine the appropriate value of $k$ to use, and note that $y_0[n]$ and $y_1[n]$ will both be of length $L+P-1$. When $y_0[n]$ and $y_1[n]$ are added together, there will generally be a region where both are nonzero. It is for this reason that this method of block convolution is called the overlap-add method. Compute $y[n]$ in this manner, and plot $y[n]$ over the range $0 \leq n \leq 99$. Is your result the same as what you found in Part ???

## Advanced Problem

(f). For this problem, you will write a MATLAB function to perform overlap-add block convolution. Your function should take as input an impulse response `h`, a data vector `x`, and a block length `L`. Your function should allow the data vector `x` to be of arbitrary length, and the block length `L` to be an arbitrary number greater than the length of the filter. The first line of your function should read

```
function y = oafilt(h,x,L)
```

Redo Part ?? using your function and verify that your function works correctly by comparing your result with that obtained from direct convolution using `conv`.

## ■ 2.8 Numerical Approximations of Continuous-Time Convolution

In this exercise, you will use MATLAB to compute numerical approximations to the convolution integral. In order to approximate the continuous functions in this integral using MATLAB vectors, you will use piecewise constant functions. The convolution of the piecewise constant functions yields an expression which converges to the convolution of the original continuous-time signals and which is simple to evaluate in MATLAB at appropriately chosen times.

Let $\delta_\Delta(t)$ be defined as a rectangular pulse of width $\Delta$ and height 1 centered on $t = 0$, i.e.,

$$\delta_\Delta(t) = \begin{cases} 1, & -\Delta/2 \leq t < \Delta/2, \\ 0, & \text{otherwise}. \end{cases}$$

A function $x(t)$ can be approximated by a piecewise constant function $x_\Delta(t)$ consisting of a sequence of pulses spaced every $\Delta$ in time with heights $x(k\Delta)$:

$$x_\Delta(t) = \sum_{k=-\infty}^{\infty} x(k\Delta)\delta_\Delta(t - k\Delta).$$

In the limit as $\Delta \to 0$, $x_\Delta(t) \to x(t)$ for a large class of functions. For example, Figure ?? shows $x_\Delta(t)$ for $x(t) = e^{-2t}u(t)$ and $\Delta = 0.25$. From this figure, it is easy to visualize how $x_\Delta(t)$ would become a better approximation to $x(t)$ as $\Delta$ is decreased.

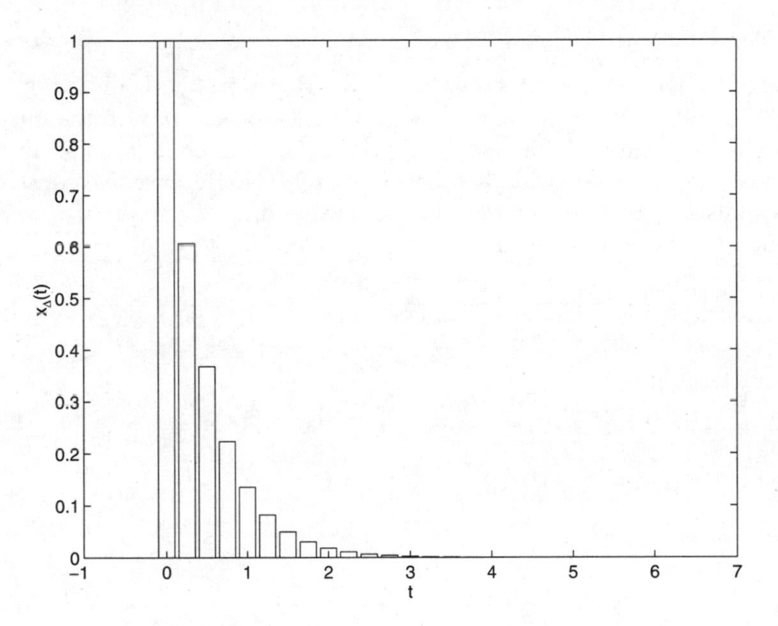


**Figure 2.11.** Piecewise constant $x_\Delta(t)$ as an approximation to $x(t) = e^{-2t}u(t)$ ($\Delta = 0.25$).

Similarly, $h(t)$ can be approximated by

$$h_\Delta(t) = \sum_{k=-\infty}^{\infty} h(k\Delta)\delta_\Delta(t - k\Delta).$$

The convolution integral for an LTI system

$$y(t) = \int_{-\infty}^{\infty} x(\tau)h(t - \tau)d\tau \tag{2.17}$$

can be approximated by convolving the piecewise constant signals

$$y_\Delta(t) = \int_{-\infty}^{\infty} x_\Delta(\tau)h_\Delta(t-\tau)d\tau.$$

Note that $y_\Delta(t)$ is not necessarily itself piecewise constant, but rather the convolution of two piecewise constant signals.

For this exercise, you will approximate the convolution of $x(t) = e^{-2t}u(t)$ with $h(t) = u(t)$ using piecewise constant functions for several different values of $\Delta$. As $\Delta \to 0$, this approximation will converge to the analytic convolution of these two functions.

## Basic Problem

(a). Find $y(t) = x(t) * h(t)$ analytically for the functions $x(t)$ and $h(t)$ defined above. You will need to know the result of this convolution so you can compare it to the numerical approximations you will be computing in this exercise.

## Intermediate Problems

In general, solving for $y_\Delta(t)$ can be complicated. However, the convolution is simple to evaluate when $t$ is an integer multiple of $\Delta$, so you will only compute the output at those times, i.e., $y_\Delta(n\Delta)$. Figure **??** shows $x_\Delta(\tau)$ and $h_\Delta(1-\tau)$ for $\Delta = 0.25$. Integrating the product of these two functions will give the value of $y_\Delta(1)$. Because the constant segments of the two signals are perfectly aligned for this value of $t$, the integral is very simple to evaluate. When $t = n\Delta$, the segments will always align this way.

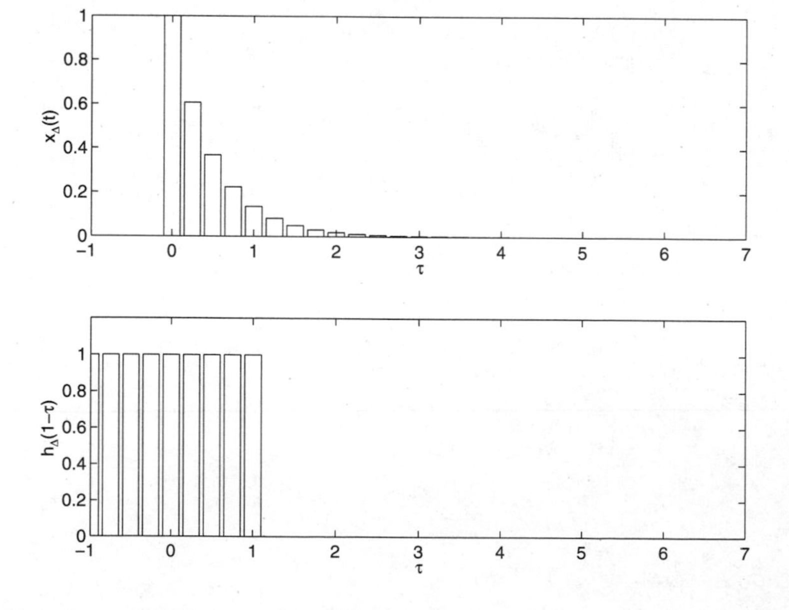


**Figure 2.12.** $x_\Delta(\tau)$ and $h_\Delta(1-\tau)$ for $\Delta = 0.25$.

(b). Starting from the convolution integral, Eq. (??), derive that, in general,

$$y_\Delta(n\Delta) = \Delta \sum_{k=-\infty}^{\infty} x(k\Delta)\, h((n-k)\Delta). \tag{2.18}$$

The expression you derived in Part ?? can be used to approximate $y(t)$. If the signals $x_\Delta(t), h_\Delta(t)$, and $y_\Delta(t)$ are represented by MATLAB vectors containing the values of the signals at times $t = n\Delta$, then Eq. (??) is a discrete convolution sum of the MATLAB vectors. By computing this discrete convolution as the value of $\Delta$ decreases, you can obtain closer approximations to $y(t)$.

(c). Define `x1` to be the values of $x_\Delta(n\Delta)$ on the interval $0 \leq t \leq 6$ with $\Delta = 0.5$. The $m$-th element of `x1` should be $x_\Delta((m-1)\Delta)$, i.e., `x(1)` $= x_\Delta(0)$. Similarly, define `h1` to be samples of $h(t)$ for $\Delta = 0.5$ over the same time interval.

(d). Tutorial ?? demonstrates how to use the `conv` function to implement a convolution sum. Use `conv` to compute `y1`, the discrete convolution of `x1` and `h1`. Don't forget to multiply the result of your convolution by $\Delta$ as shown in Eq. (??). Note that because `x1` and `h1` are finite-length segments of infinite-length signals, only a finite-length segment of `y1` is a valid approximation to $y(t)$. The rest of `y1` must be discarded. For what values of $t$ does `y1` approximate $y(t)$? Plot the valid elements of `y1` with a correctly labeled time axis. Is `y1` a close approximation to the analytic expression you found in Part ???

(e). Represent $x_\Delta(n\Delta)$ for $\Delta = 0.25$, 0.1, and 0.01 on the time interval $0 \leq t \leq 6$ using vectors `x2` through `x4`. Define `h2` through `h4` to represent $h_\Delta(n\Delta)$ on the same interval with the same sampling periods. Compute `y2` through `y4` using `conv` and the appropriate inputs and impulse responses. In each case, the output signal is valid over the same range of $t$, but because $\Delta$ is different for each output, this corresponds to a different set of samples. Plot `y1` through `y4` on the same set of axes. For which values of $\Delta$ is the asymptotic value of the approximation within 10% of the asymptotic value of the true $y(t)$ that you found in Part ???

## ■ 2.9 The Pulse Response of Continuous-Time LTI Systems

An important property of an LTI system is that it is completely defined by its impulse response[2]. If $h(t)$ is the response of a continuous-time LTI system to the unit impulse $\delta(t)$, then

$$y(t) = \int_{-\infty}^{\infty} h(t-\tau)\, x(\tau)\, d\tau$$

is the system response to any input $x(t)$. However, LTI systems also can be characterized effectively by their response to a pulse of finite duration as long as the duration of the pulse

[2] A more complete discussion of $\delta(t)$ and its role in the characterization of LTI systems is given in Section 2.5 of *Signals and Systems*.

is chosen to be short enough. Toward that end, define $\delta^{\Delta}(t)$ to be the pulse

$$\delta^{\Delta}(t) = \begin{cases} \frac{1}{\Delta}, & 0 \le t < \Delta, \\ 0, & \text{otherwise}. \end{cases} \tag{2.19}$$

Note that the area of this pulse remains constant and equal to one for all values of $\Delta > 0$. If $h^{\Delta}(t)$ is the response of an LTI system to the pulse $\delta^{\Delta}(t)$, and

$$y^{\Delta}(t) = \int_{-\infty}^{\infty} h^{\Delta}(t-\tau)\, x(\tau)\, d\tau,$$

then $y^{\Delta}(t) \approx y(t)$ if $\Delta$ is chosen small enough. By "small enough," we mean that the difference between $h(t)$ and $h^{\Delta}(t)$ is of no practical significance. The values for which $\Delta$ is small enough will be a function of the particular LTI system. However, for a large class of signals, we have that $h^{\Delta}(t) \to h(t)$ as $\Delta \to 0$.
For this exercise, you will consider the causal LTI system whose input $x(t)$ and output $y(t)$ satisfy the following first-order linear differential equation:

$$\frac{dy(t)}{dt} + 3\,y(t) = x(t). \tag{2.20}$$

You will consider the response of this system to pulses of various shapes and durations, and will discover how accurately $h^{\Delta}(t)$ characterizes the system for various values of $\Delta$.

## Basic Problems

(a). Analytically derive the unit step response $s(t)$ and the unit impulse response $h(t)$ of the causal LTI system defined by Eq. (??). Hint: The impulse response is equal to $ds(t)/dt$. Store in `h` and `s` the values of $h(t)$ and $s(t)$ at each time sample in the vector `t=[-1:0.05:4]`. Plot both `h` and `s` versus `t`.

(b). Use `step` and `impulse` to verify the functions that you computed in Part ??. The functions `step(b,a,t)` and `impulse(b,a,t)` are explained in Tutorial ??, the tutorial for `lsim`. The vectors `b` and `a` are determined by the coefficients in Eq. (??). For the time vector `t`, use `t=[0:0.05:4]`.

## Intermediate Problems

In Tutorial ??, you learned how to use `lsim` to determine the response of a causal continuous-time LTI system described by a linear constant-coefficient differential equation. You will now use `lsim` to simulate the response $h^{\Delta}(t)$ to pulses of different shapes and durations.

(c). Use `lsim` to simulate the response to $\delta^{\Delta}(t)$ for $\Delta = 0.1$, 0.2, and 0.4. For the time vector, use `t=[-1:0.05:4]`, and store in `d1`, `d2`, and `d3` the corresponding values of $\delta^{\Delta}(t)$ for $\Delta = 0.1$, 0.2, and 0.4, respectively. Remember that $\delta^{\Delta}(\Delta) = 0$, not $1/\Delta$. Store in `h1`, `h2`, and `h3` the simulated response for $\Delta = 0.1$, 0.2, and 0.4, respectively. These vectors contain the simulated values of $h^{\Delta}(t)$ at the time samples given in `t`.

On three separate figures, plot the simulated values of $h^{\Delta}(t)$. On each figure, also include a plot of $h(t)$, which was stored in `h` in Part ??. How does $h^{\Delta}(t)$ compare to $h(t)$ as $\Delta$ decreases?

(d). Why is it necessary to ensure that the pulse $\delta^{\Delta}(t)$ has unit area? In other words, what if the response to

$$D^{\Delta}(t) = \begin{cases} 1, & 0 \leq t < \Delta, \\ 0, & \text{otherwise}, \end{cases}$$

were used instead as the approximation of $h(t)$?

(e). Now consider the function

$$d_a(t) = a\, e^{-at}\, u(t)$$

for $a > 0$. While $d_a(t)$ has infinite duration, most of the signal energy is contained in the interval $0 \leq t \leq 4/a$. For sufficiently large values of $a$, the response of the LTI system to $d_a(t)$ will for all practical purposes be identical to $h(t)$. What is the area of $d_a(t)$ as a function of $a > 0$?

Use `t=[-1:0.05:4]`, and store in `da1`, `da2`, and `da3` the corresponding values of $d_a(t)$ for $a = 4$, 8, and 16, respectively. For each of the three values of $a$, use `lsim` to simulate the response of the LTI system to $d_a(t)$. Plot each function on a separate figure using `plot`. On each figure, also include a plot of the impulse response computed in Part ??. How does the response to $d_a(t)$ compare to $h(t)$ for large values of $a$?

## Advanced Problems

These advanced problems assume that you have done Exercise ??. In Exercise ??, you learned how the convolution of two signals $y(t) = x(t) * h(t)$, could be approximated by the convolution of two piecewise constant functions. (Note that the functions $h^{\Delta}(t)$ and $h_{\Delta}(t)$, defined in this exercise and Exercise ??, respectively, have different meanings.) You can use the vector `h1` computed in Part ?? of this exercise to obtain a piecewise constant approximation of $h^{\Delta}(t)$ for $\Delta = 0.1$. Namely, since `h1(n+1)` contains the simulated values of $h^{\Delta}(t)$ at $t = 0.05\,n$, the piecewise constant approximation of $h^{\Delta}(t)$ is equal to `h1(n+1)` on the interval $0.05\,(n - 1/2) \leq t < 0.05\,(n + 1/2)$. If `x` contains the values of the piecewise constant approximation of $x(t)$ on the same intervals, then, as shown in Exercise ??, you can use `Delta*conv(h1,x)`, where `Delta` is equal to the appropriate value of $\Delta$, to approximate the values of $y(t)$ at $t = 0.05\,n$.
In these problems, you seek to show that $h^{\Delta}(t)$ for $\Delta = 0.1$ can effectively represent the LTI system given by Eq. (??).

(f). For the time vector `t`, use `t=[-1:0.05:4]`, and store in the vector `u` the values of the unit step function $u(t)$ evaluated at each time in `t`.

(g). The function $h^{\Delta}(t)$ can be used to compute an approximation of the unit step response as $s^{\Delta}(t) = h^{\Delta}(t) * u(t)$. Use `conv(h1,u)` to approximate $s^{\Delta}(t)$. Remember that the discrete convolution must be scaled by $\Delta$, and that only a portion of the signal returned by `conv(h1,u)` contains samples of $s^{\Delta}(t)$. Plot the approximate step response computed with `conv` on the same figure as the samples of the exact step response computed in Part ?? and stored in the vector `s`. How will the fidelity of the discrete-time approximation to the step response change as $\Delta$ increases? Explain.

## 2.10 Echo Cancellation via Inverse Filtering

In this exercise, you will consider the problem of removing an echo from a recording of a speech signal. This project will use the audio capabilities of MATLAB to play recordings of both the original speech and the result of your processing. To begin this exercise you will need to load the speech file `lineup.mat`, which is contained in the Computer Explorations Toolbox. The Computer Explorations Toolbox can be obtained from The MathWorks at the address provided in the Preface. If this speech file is already somewhere in your MATLABPATH, then you can load the data into MATLAB by typing

```
>> load lineup.mat
```

You can check your MATLABPATH, which is a list of all the directories which are currently accessible by MATLAB, by typing `path`. The file `lineup.mat` must be in one of these directories.

Once you have loaded the data into MATLAB, the speech waveform will be stored in the variable `y`. Since the speech was recorded with a sampling rate of 8192 Hz, you can hear the speech by typing

```
>> sound(y,8192)
```

You should hear the phrase "line up" with an echo. The signal $y[n]$, represented by the vector `y`, is of the form

$$y[n] = x[n] + \alpha x[n-N], \tag{2.21}$$

where $x[n]$ is the uncorrupted speech signal, which has been delayed by $N$ samples and added back in with its amplitude decreased by $\alpha < 1$. This is a reasonable model for an echo resulting from the signal reflecting off of an absorbing surface like a wall. If a microphone is placed in the center of a room, and a person is speaking at one end of the room, the recording will contain the speech which travels directly to the microphone, as well as an echo which traveled across the room, reflected off of the far wall, and then into the microphone. Since the echo must travel further, it will be delayed in time. Also, since the speech is partially absorbed by the wall, it will be decreased in amplitude. For simplicity ignore any further reflections or other sources of echo.

For the problems in this exercise, you will use the value of the echo time, $N = 1000$, and the echo amplitude, $\alpha = 0.5$.

## Basic Problems

(a). In this exercise you will remove the echo by linear filtering. Since the echo can be represented by a linear system of the form Eq. (??), determine and plot the impulse response of the echo system Eq. (??). Store this impulse response in the vector `he` for $0 \leq n \leq 1000$.

(b). Consider an echo removal system described by the difference equation

$$z[n] + \alpha z[n - N] = y[n], \tag{2.22}$$

where $y[n]$ is the input and $z[n]$ is the output which has the echo removed. Show that Eq. (??) is indeed an inverse of Eq. (??) by deriving the overall difference equation relating $z[n]$ to $x[n]$. Is $z[n] = x[n]$ a valid solution to the overall difference equation?

## Intermediate Problems

(c). The echo removal system Eq. (??) will have an infinite-length impulse response. Assuming that $N = 1000$, and $\alpha = 0.5$, compute the impulse response using `filter` with an input that is an impulse given by `d=[1 zeros(1,4000)]`. Store this 4001 sample approximation to the impulse response in the vector `her`.

(d). Implement the echo removal system using `z=filter(1,a,y)`, where `a` is the appropriate coefficient vector derived from Eq. (??). Plot the output using `plot`. Also, listen to the output using `sound`. You should no longer hear the echo.

(e). Calculate the overall impulse response of the cascaded echo system, Eq. (??), and echo removal system, Eq. (??), by convolving `he` with `her` and store the result in `hoa`. Plot the overall impulse response. You should notice that the result is not a unit impulse. Given that you have computed `her` to be the inverse of `he`, why is this the case?

## Advanced Problem

(f). Suppose that you were given $y[n]$ but did not know the value of the echo time, $N$, or the amplitude of the echo, $\alpha$. Based on Eq. (??), can you determine a method of estimating these values? Hint: Consider the output `y` of the echo system to be of the form

$$y[n] = x[n] * (\delta[n] + \alpha\delta[n - N])$$

and consider the signal

$$R_{yy}[n] = y[n] * y[-n].$$

This is called the autocorrelation of the signal $y[n]$ and is often used in applications of echo-time estimation. Write $R_{yy}[n]$ in terms of $R_{xx}[n]$ and also plot $R_{yy}[n]$. Also try experimenting with simple echo problems such as

```
>> NX=100;
>> x=randn(1,NX);
>> N=50;
>> alpha=0.9;
>> y=filter([1 zeros(1,N-1) alpha],1,x);
>> Ryy=conv(y,fliplr(y));
>> plot([-NX+1:NX-1],Ryy)
```

by varying `N,alpha`, and `NX`. Also, when you loaded `lineup.mat`, you loaded in two additional vectors. The vector `y2` contains the phrase "line up" with a different echo time N and different echo amplitude $\alpha$. The vector `y3` contains the same phrase with two echoes, each with different times and amplitudes. Can you estimate $N$ and $\alpha$ for `y2`, and $N_1, \alpha_1, N_2$, and $\alpha_2$ for `y3`? Note that getting accurate answers for this problem is very difficult.

Chapter 3

# Fourier Series Representation of Periodic Signals

Harmonically related complex exponentials are signals whose frequencies are integer multiples of one another. The continuous-time and discrete-time Fourier series represent periodic signals as weighted sums of harmonically-related complex exponentials. These representations allow straightforward calculation of the output of a system for a given periodic input since complex exponentials are eigenfunctions of LTI systems. The exercises in this chapter allow you to explore analyzing and synthesizing periodic signals with complex exponentials, as well as how these signals behave when processed by LTI systems.
Tutorial ?? describes how to use the function `fft` to compute the discrete-time Fourier series (DTFS), while Tutorial ?? explains how to find the frequency response of a linear, constant-coefficient difference equation using `freqz`. Tutorial ?? expands on Tutorial ??, demonstrating that `lsim` can be used to simulate a continuous-time system from its system function as well as from its differential equation. Exercises ?? and ?? explore the eigenfunction property and synthesis of discrete-time signals. The properties of the continuous-time Fourier series (CTFS) are demonstrated in Exercises ?? and ??. Exercises ?? and ?? study the effect of simple filters on periodic discrete-time and continuous-time signals. The computational efficiency of `fft` over a straight-forward implementation of the DTFS is demonstrated in Exercise ??.
This chapter also contains several exercises that use the MATLAB Symbolic Math Toolbox to study periodic signals. Exercise ?? synthesizes several continuous-time signals from the CTFS using different choices for the period $T$, while Exercise ?? focuses exclusively on square and triangle waves. Exercise ?? simulates the response of a simple RL circuit to periodic inputs.

## ■ 3.1 Tutorial: Computing the Discrete-Time Fourier Series with `fft`

The discrete-time Fourier series (DTFS) is a frequency-domain representation for periodic discrete-time sequences. For a signal $x[n]$ with fundamental period $N$, the DTFS synthesis and analysis equations are given by

$$x[n] = \sum_{k=0}^{N-1} a_k \, e^{jk(2\pi/N)n} \tag{3.1}$$

and

$$a_k = \frac{1}{N} \sum_{n=0}^{N-1} x[n] \, e^{-jk(2\pi/N)n} , \tag{3.2}$$

respectively. Remember that $x[n]$ has period $N$, so that the summation in Eq. (??) can be replaced with a sum over any $N$ consecutive values of $n$. Similarly, $a_k$ is periodic in $k$ with period $N$ so that the summation in Eq. (??) can be replaced with a sum over any $N$ consecutive values of $k$.

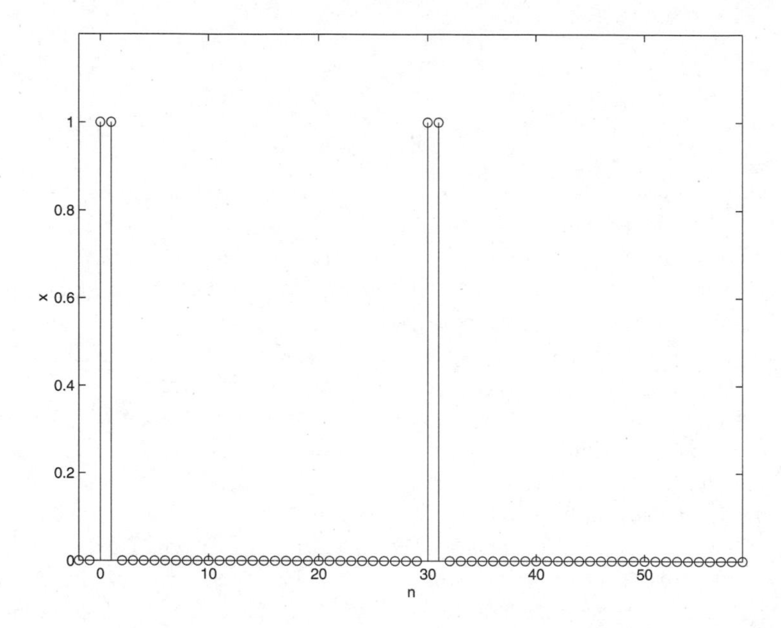


**Figure 3.1.** Periodic discrete-time signal $x[n]$ with fundamental period $N = 30$.

MATLAB contains two very efficient routines for computing Eqs. (??) and (??). If `x` is an $N$-point vector containing $x[n]$ for the single period $0 \le n \le N-1$, then the DTFS of $x[n]$ can be computed by `a=(1/N)*fft(x)`, where the $N$-point vector `a` contains $a_k$ for $0 \le k \le N-1$. The function `fft` is simply an efficient implementation of Eq. (??) scaled by $N$. For example, assume $x[n]$ is the signal with fundamental period $N = 30$ plotted in Figure ??. The signal is given by

$$x[n] = \begin{cases} 1, & n = 0, 1, \\ 0, & \text{otherwise} \end{cases}$$

on the interval $0 \le n \le 29$. Define `x=[1 1 zeros(1,28)]`. The DTFS can be computed by typing `a=(1/N)*fft(x)`. The real and imaginary parts of `a` are plotted in Figures ?? and ??. You can verify analytically that these are the correct values for $a_k$.
Given a vector `a` containing the DTFS coefficients $a_k$ for $0 \le k \le N-1$, the function `ifft` can be used to construct a vector `x` containing $x[n]$ for $0 \le n \le N-1$ as follows:

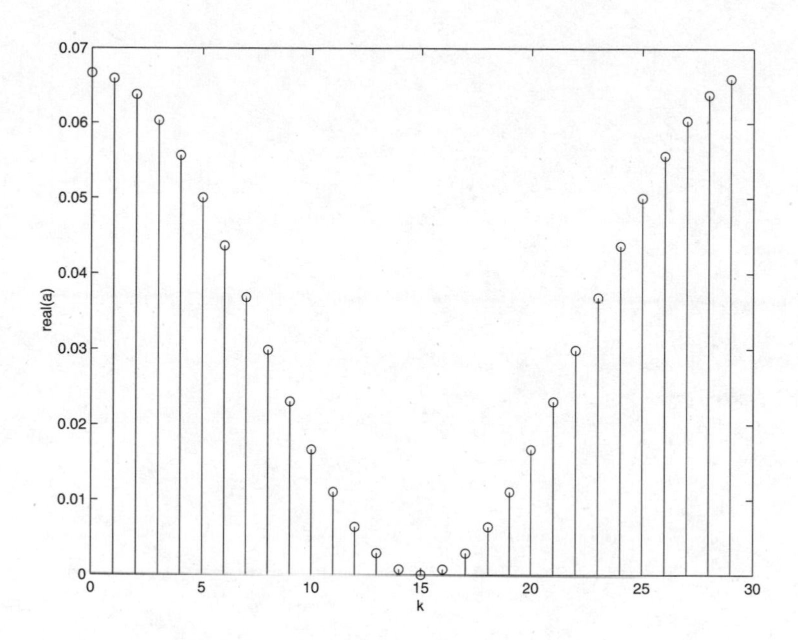


**Figure 3.2.** The real part of $a_k$ for $0 \leq k \leq N-1$.

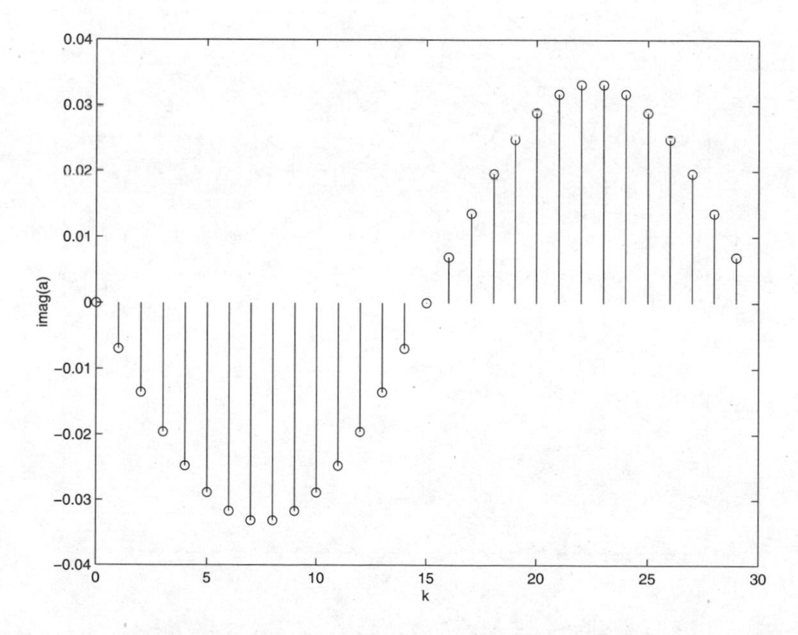


**Figure 3.3.** The imaginary part of $a_k$ for $0 \leq k \leq N-1$.

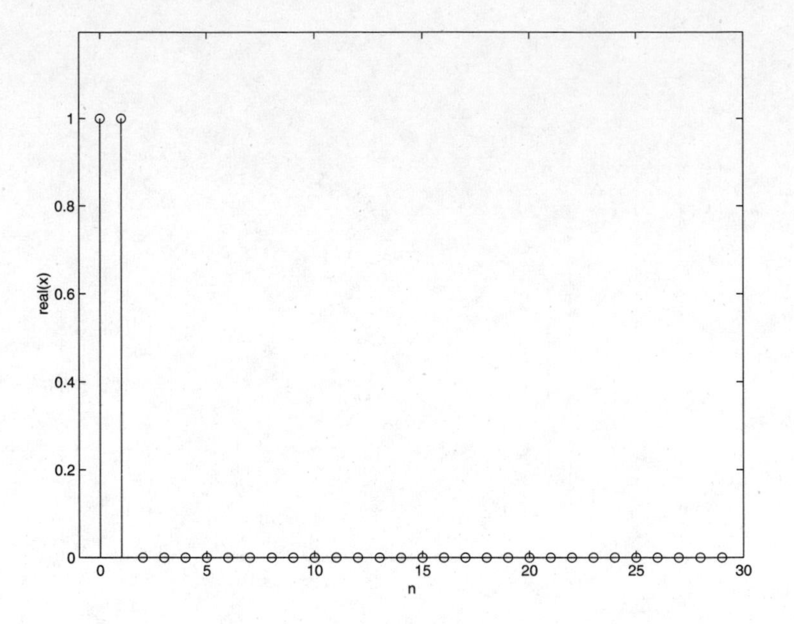


**Figure 3.4.** Real part of the synthesized discrete-time signal.

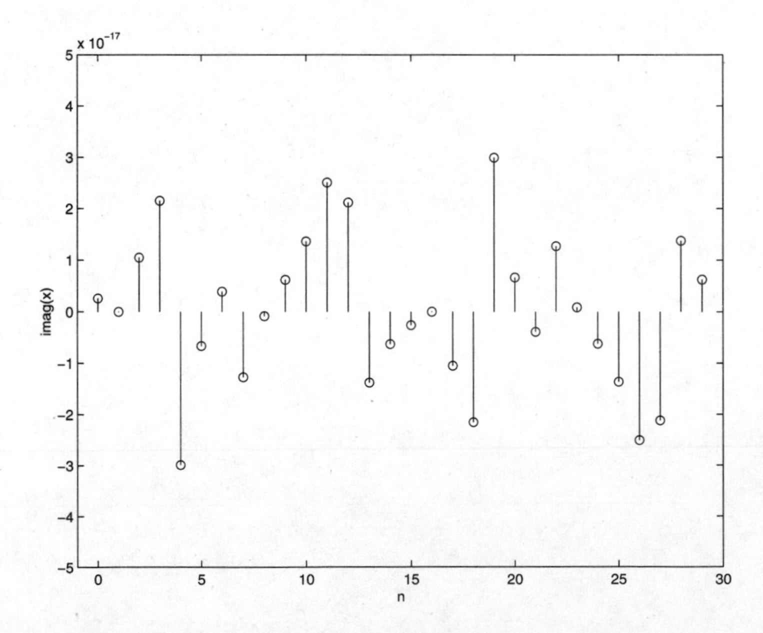


**Figure 3.5.** Imaginary part of the synthesized discrete-time signal. Note that the imaginary component is insignificant compared to the real component in Figure ??.

x=N*ifft(a). The function ifft is an efficient realization of the DTFS synthesis equation, scaled by $1/N$. From the vector a whose real and imaginary parts are plotted in Figures ??-??, we obtain a signal N*ifft(a) whose real and imaginary parts are plotted in Figures ?? and ??. While the signal plotted in Figure ?? had no imaginary component, the imaginary component of N*ifft(a), as shown in Figure ??, is not identically zero. However, close inspection of the ordinate of Figure ?? reveals that the imaginary component is very small — within $\pm 10^{-16}$. This imaginary component is due to numerical round-off errors. Numerical round-off errors result from the finite-precision arithmetic used by any computer implementation. For the exercises in this book, numerical round-off errors are generally insignificant and can be ignored. For this signal, N*ifft(a) should be real, so you would use real(N*ifft(a)) to synthesize the signal $x[n]$.
In essence, fft and ifft implement the equations

$$x[n] = \frac{1}{N} \sum_{k=0}^{N-1} a_k \, e^{jk(2\pi/N)n}$$

and

$$a_k = \sum_{n=0}^{N-1} x[n] \, e^{-jk(2\pi/N)n} \,,$$

respectively. This pair of synthesis and analysis equations is a valid alternative to Eqs. (??) and (??), and is a convention adopted by some texts. The exercises in this book will follow the convention of *Signals and Systems*, i.e., Eqs. (??) and (??).

## ■ 3.2 Tutorial: freqz

The signals $e^{j\omega n}$ are eigenfunctions of LTI systems. For each value of $\omega$ the frequency response $H(e^{j\omega})$ is the eigenvalue of the LTI system for the eigenfunction $e^{j\omega n}$; when the input sequence is $x[n] = e^{j\omega_0 n}$, the output sequence is $y[n] = H(e^{j\omega_0})e^{j\omega_0 n}$. For a causal LTI system described by a difference equation, the command [H omega]=freqz(b,a,N) computes the frequency response $H(e^{j\omega})$ at $N$ evenly spaced frequencies between 0 and $\pi$, i.e., $\omega_k = (\pi/N)k$ for $0 \le k \le N-1$. The coefficient vectors a and b specify the difference equation using the same format as Eq. (??) in the filter tutorial. For the command above, freqz returns $H(e^{j\omega_k})$ in H and the frequencies $\omega_k$ in omega. Including the 'whole' option as [H omega]=freqz(b,a,N,'whole') computes the samples of the frequency response $H(e^{j\omega})$ at $N$ evenly spaced frequencies from 0 to $2\pi$, $\omega_k = (2\pi/N)k$ for $0 \le k \le N-1$.

(a). Define a1 and b1 to describe the causal LTI system specified by the difference equation $y[n] - 0.8y[n-1] = 2x[n] - x[n-2]$.

(b). Use freqz with the coefficients from Part ?? to define H1 to be the value of the frequency response at 4 evenly spaced frequencies between 0 and $\pi$ and omega1 to be those frequencies. The following sample output shows the values each vector should have if you have defined things correctly:

```
>> H1.'
ans =
   5.0000     2.8200 - 1.3705i   1.8293 - 1.4634i   0.9258 - 0.9732i
>> omega1.'
ans =
         0    0.7854    1.5708    2.3562
```

(c). Use `freqz` to define `H2` to be the value of the frequency response at 4 evenly spaced frequencies between 0 and $2\pi$ and `omega2` to be those frequencies. The following sample output shows the values each vector should have if you have defined things correctly:

```
>> H2.'
ans =
   5.0000     1.8293 - 1.4634i   0.5556    1.8293 + 1.4634i
>> omega2.'
ans =
         0    1.5708    3.1416    4.7124
```

## 3.3 Tutorial: `lsim` with System Functions

Tutorial ?? describes how the `lsim` command can be used to simulate a causal LTI continuous-time system whose input and output satisfy a linear constant-coefficient differential equation. The output of a causal LTI system specified by its system function can also be simulated using `lsim`, since the system function uniquely specifies the differential equation relating the input and output of the system. If the system function is put in the form

$$H(s) = \frac{\mathtt{b}(1)s^M + \ldots + \mathtt{b}(M-1)s + \mathtt{b}(M)}{\mathtt{a}(1)s^N + \ldots + \mathtt{a}(N-1)s + \mathtt{a}(N)},$$

the output of the system for an input $x(t)$ can be simulated using `lsim(b,a,x,t)`, where the MATLAB vectors `b` and `a` contain the coefficients of the numerator and denominator polynomials in $s$, and the vectors `x` and `t` describe the input signal in the same format specified in Tutorial ??. Note that $H(s)$ must be a proper fraction, i.e., $N \geq M$.
As an example, consider the system function

$$H(s) = \frac{s + \frac{1}{2}}{s - 2},$$

whose coefficients are defined by the vectors `b=[1 1/2]` and `a=[1 -2]`. The command `y=lsim(b,a,x,t)` stores in `y` the time response of the system to the input specified in the vector `x` at the times specified in `t`. The vector `y` has as many elements as the input vector `x`.

(a). Define coefficient vectors `a1` and `b1` to describe the causal LTI system specified by the system function

$$H_1(s) = \frac{s - 2}{s + 2}.$$

(b). Define coefficient vectors a2 and b2 to describe the causal LTI system specified by the system function

$$H_2(s) = \frac{3}{s+0.3}.$$

(c). Define coefficient vectors a3 and b3 to describe the causal LTI system specified by the system function

$$H_3(s) = \frac{2s}{s+0.8}.$$

(d). Use `lsim` and vectors you defined in the previous parts to find the output of those causal LTI systems for the input given by `t=[0:0.1:0.5]`, `x=cos(t)`. The results shown below give the output for each system:

```
>> y1=lsim(b1,a1,x,t)'
y1 =
        1.0000    0.6334    0.3261    0.0692   -0.1444   -0.3205

>> y2=lsim(b2,a2,x,t)'
y2 =
        0    0.2948    0.5779    0.8468    1.0991    1.3323

>> y3=lsim(b3,a3,x,t)'
y3 =
        2.0000    1.8366    1.6667    1.4910    1.3105    1.1262
```

## 3.4 Eigenfunctions of Discrete-Time LTI Systems

This exercise examines the eigenfunction property of discrete-time LTI systems. Complex exponentials are eigenfunctions of LTI systems, i.e., when the input sequence is a complex exponential, the output is the same complex exponential only scaled in amplitude by a complex constant. This constant can be computed from the impulse response $h[n]$. When the input to a discrete-time LTI system is $x[n] = z^n$, the output is $y[n] = H(z)z^n$, where

$$H(z) = \sum_{n=-\infty}^{\infty} h[n]z^{-n}.$$

Consider each of the following input signals:

$$\begin{aligned} x_1[n] &= e^{j(\pi/4)n}, \\ x_2[n] &= \sin(\pi n/8 + \pi/16), \\ x_3[n] &= (9/10)^n, \\ x_4[n] &= n+1. \end{aligned}$$

You will compute the outputs $y_1[n]$ through $y_4[n]$ that result when each of these signals is the input to the causal LTI system described by the linear, constant-coefficient difference equation

$$y[n] - 0.25y[n-1] = x[n] + 0.9x[n-1]. \tag{3.3}$$

## Basic Problems

(a). Create a vector `n` containing the time indices for the interval $-20 \leq n \leq 100$ using the colon (`:`) operator. Using this vector, define `x1`, `x2`, `x3`, and `x4` to be vectors containing the values of the four signals $x_1[n]$ through $x_4[n]$ over the interval described by `n`. Produce a clearly labeled plot of each signal over this interval. Since the vector `x1` is complex, you will need to produce two separate plots for the real and imaginary parts. You can combine these in a single figure using either `subplot` or `hold`.

(b). The `filter` command described in Tutorial **??** computes the output of the causal, LTI system described by a difference equation for a given input sequence. Define `a` and `b` to specify the system shown in Eq. (**??**). Use these vectors and `filter` to compute the vectors `y1`, `y2`, `y3` and `y4`, containing the output of the system specified by Eq. (**??**) when the input is `x1` through `x4`, respectively. For each of the outputs you obtain, produce appropriately labeled plots of the portion of the outputs over the interval $0 \leq n \leq 100$. For `y1`, you will again need to plot the real and imaginary parts separately. Comparing your plots of inputs and outputs, indicate which of the input signals are eigenfunctions of this LTI system.

Note: In both this part and the next part we will ignore the output samples over the interval $-20 \leq n \leq -1$. These samples include transients due to the natural response of the system because MATLAB is unable to work with infinitely long input signals. MATLAB assumes the input and output were zero before the values given in `x`. The eigenfunction property of the system relates only to the steady-state solution. The signals and systems in this exercise have been chosen to insure the transients have died out completely within 20 samples, so by restricting the interval examined to be $0 \leq n \leq 100$, you are working with a portion of the output where the effect of the natural response is insignificant.

(c). For this part, you will verify which of the inputs were eigenfunctions and compute the corresponding eigenvalues for those eigenfunctions. If the vectors `x` and `y` describe the input and output sequences of a system, and the input sequence is an eigenfunction of the system, `y` should be equal to `x` scaled by a constant. This can be verified by computing `H=y./x`, which computes the ratio of the output to input sequence at each time index. If the resulting vector `H` is a constant, the input signal was an eigenfunction of the system. Compute `H1` through `H4` for each of the input/output signal pairs you obtained above, and produce appropriately labeled plots of `H` over the interval $0 \leq n \leq 100$ again. Again, `H1` will require separate plots for its real and imaginary parts. For the inputs which are eigenfunctions of the system, find the eigenvalue $H(z)$ from your plots or from the vector `H`.

## 3.5 Synthesizing Signals with the Discrete-Time Fourier Series

The discrete-time Fourier series (DTFS) is a frequency-domain representation for periodic discrete-time sequences. The synthesis and analysis equations for the DTFS are given by Eqs. (??) and (??). This exercise contains three sets of problems to give you practice working with these equations. The Basic Problems allow you to synthesize a very simple periodic discrete-time signal from its DTFS coefficients. In the Intermediate Problems, you will both analyze a set of periodic discrete-time signals to obtain their DTFS coefficients, and construct one of these signals by adding in a few coefficients at a time. For the Advanced Problem, you will write a function to find the DTFS coefficients of an arbitrary periodic discrete-time signal from one period of samples.

### Basic Problems

In these problems, you will synthesize a periodic discrete-time signal with period $N = 5$ and the following DTFS coefficients:

$$a_0 = 1,\ a_2 = a_{-2}^* = e^{j\pi/4},\ a_4 = a_{-4}^* = 2e^{j\pi/3}.$$

(a). Based on the DTFS coefficients, do you expect $x[n]$ to be complex-valued, purely real, or purely imaginary? Why?

(b). Using the DTFS coefficients given above, determine the values of $a_0$ through $a_4$ and specify a vector `a` containing these values.

(c). Using the vector `a` of DTFS coefficients and the synthesis equation, define a new vector `x` containing one period of the signal $x[n]$ for $0 \leq n \leq 4$. You can either write out the summation explicitly or you may find it helpful to use a `for` loop. Generate an appropriately labeled plot of $x[n]$ for $0 \leq n \leq 4$ using `stem`. Was your prediction in Part ?? correct? Note that if you predicted a purely imaginary or real signal, it may still have a very small ($< 10^{-10}$) nonzero real or imaginary part due to roundoff errors. If this is the case, set this part to be zero using `real` or `imag` as appropriate before making your plot.

### Intermediate Problems

For these problems, you will examine the DTFS representation of several different square waves. Specifically, you will look at signals

$$x_1[n] = \begin{cases} 1, & 0 \leq n \leq 7, \end{cases} \tag{3.4}$$

$$x_2[n] = \begin{cases} 1, & 0 \leq n \leq 7, \\ 0, & 8 \leq n \leq 15, \end{cases} \tag{3.5}$$

$$x_3[n] = \begin{cases} 1, & 0 \leq n \leq 7, \\ 0, & 8 \leq n \leq 31, \end{cases} \tag{3.6}$$

where $x_1[n]$, $x_2[n]$ and $x_3[n]$ have periods of $N_1 = 8$, $N_2 = 16$ and $N_3 = 32$, respectively.

(d). Define three vectors x1, x2, and x3 representing one period of each of the signals $x_1[n]$, $x_2[n]$ and $x_3[n]$. Using these vectors, make appropriately labeled plots of each of the signals over the range $0 \leq n \leq 63$. Note: You will have to repeat the vectors you have defined to cover this range of samples.

(e). Exercise ?? explains how to use fft to compute the DTFS of a periodic discrete-time signal from one period of the signal. Using the fft function, define the vectors a1, a2, and a3 to be the DTFS coefficients of $x_1[n]$ through $x_3[n]$, respectively. Generate appropriately labeled plots of the magnitude of each of the DTFS coefficient sequences using abs and stem. From your time domain plots of Part ?? and Eq. (??), you should be able to predict the values of a1(1), a2(1) and a3(1)—the DC components of the signals. Do your predicted values match those obtained with MATLAB?

(f). In this part, you will examine how $x_3[n]$ appears when it is synthesized a few coefficients at a time. Using the vector a3 you found in the previous part, define vectors x3_2, x3_8, x3_12 and x3_all corresponding to the four signals

$$x_{3\_2}[n] = \sum_{k=-2}^{2} a_k e^{jk(2\pi/32)n},$$

$$x_{3\_8}[n] = \sum_{k=-8}^{8} a_k e^{jk(2\pi/32)n},$$

$$x_{3\_12}[n] = \sum_{k=-12}^{12} a_k e^{jk(2\pi/32)n},$$

$$x_{3\_\text{all}}[n] = \sum_{k=-15}^{16} a_k e^{jk(2\pi/32)n},$$

on the interval $0 \leq n \leq 31$. Note that since $x_3[n]$ is real, the DTFS coefficients for this signal will be conjugate symmetric, i.e., $a_k = a^*_{-k}$. Because $a_k$ is conjugate symmetric and all of the sums except $x_{3\_\text{all}}[n]$ are symmetric about $k = 0$, the resulting time signals should be purely real. If you are unclear about why this is true, you may want to review the symmetry properties of the DTFS. Due to roundoff error in MATLAB, you may need to discard some very small imaginary parts of the signals you synthesize using real. The sums specified above are symmetric about $k = 0$ but the vector a3 represents $a_k$ for $k = 0, \ldots, 31$ as [a3(1), ..., a3(32)], so you will need to determine which elements of a correspond to the negative values of $k$ when you implement the sums.

(g). Argue that $x_{3\_\text{all}}[n]$ must be a real signal.

(h). Generate a sequence of appropriately labeled plots using stem showing how the signals you created converge to $x_3[n]$ as more of the DTFS coefficients are included in the sum. Specifically, x3_all should be equal to the original vector x3 within the bounds of MATLAB's roundoff error. Does the synthesis of this discrete-time square wave display the Gibb's phenomenon?

## Advanced Problem

For this problem, you will write a function which computes the DTFS coefficients of a periodic signal. Your function should take as arguments the vector `x`, which specifies the values of the signal $x[n]$ over one period, and `n_init` which specifies the time index $n$ of the first sample of `x`. Your function should return the vector `a` containing the DTFS coefficients $a_0$ through $a_{N-1}$, where $N$ is the number of samples in `x`, or equivalently, the period of $x[n]$. The first line of your M-file should read

```
function a = dtfs(x,n_init)
```

Verify that your function is working correctly by computing the DTFS coefficients for the signals given below, and demonstrate that the output of your function matches the outputs below to within machine precision:

```
>> dtfs([1 2 3 4],0)
ans =
   2.5000            -0.5000 + 0.5000i  -0.5000              -0.5000 - 0.5000i
>> dtfs([1 2 3 4],1)
ans =
   2.5000             0.5000 + 0.5000i   0.5000 + 0.0000i   0.5000 - 0.5000i
>> dtfs([1 2 3 4],-1)
ans =
   2.5000            -0.5000 - 0.5000i   0.5000 - 0.0000i  -0.5000 + 0.5000i
>> dtfs([2 3 4 1],0)
ans =
   2.5000            -0.5000 - 0.5000i   0.5000              -0.5000 + 0.5000i
>> dtfs([ones(1,4) zeros(1,4)],0)
ans =
  Columns 1 through 4
   0.5000             0.1250 - 0.3018i        0             0.1250 - 0.0518i
  Columns 5 through 8
        0             0.1250 + 0.0518i        0             0.1250 + 0.3018i
>> dtfs([ones(1,4) zeros(1,4)],2)
ans =
  Columns 1 through 4
   0.5000            -0.3018 - 0.1250i        0             0.0518 + 0.1250i
  Columns 5 through 8
        0             0.0518 - 0.1250i        0            -0.3018 + 0.1250i
```

## ■ 3.6 Properties of the Continuous-Time Fourier Series

This exercise examines properties of the continuous-time Fourier series (CTFS) representation for periodic continuous-time signals. Consider the signal

$$x_1(t) = \cos(\omega_0 t) + \sin(2\omega_0 t), \tag{3.7}$$

where $\omega_0 = 2\pi$. To evaluate this signal in MATLAB, use the time vector

```
>> t=linspace(-1,1,1000);
```

which creates a vector of 1000 time samples over the region $-1 \le t \le 1$.

### Intermediate Problems

(a). What is the smallest period, $T$, for which $x_1(t) = x_1(t+T)$? Analytically find the coefficients of the CTFS for $x_1(t)$ using this value of $T$.

(b). Consider the signal $y(t) = x_1(t) + x_1(-t)$. Using the time-reversal and conjugation properties of the CTFS, determine the coefficients for the CTFS of $y(t)$.

(c). Plot the signal $y(t)$ over $-1 \le t \le 1$. What type of symmetry do you expect? Can you explain what you see in terms of the symmetry properties of the CTFS?

(d). Consider the signal $z(t) = x_1(t) - x_1^*(-t)$. Using the time-reversal and conjugation properties of the CTFS, determine the coefficients for the CTFS of $z(t)$.

(e). Plot the signal $z(t)$ over $-1 \le t \le 1$. What type of symmetry do you expect to see? Can you explain what you see in terms of the symmetry properties of the CTFS?

(f). Repeat Parts ??–?? using

$$x_2(t) = \cos(\omega_0 t) + i\sin(2\omega_0 t). \tag{3.8}$$

Note that $x_2(t)$ is complex. When plotting $x_2(t)$, $y(t)$, and $z(t)$, be sure to plot the real and imaginary parts separately and take note of the symmetry that you see in each.

## ■ 3.7 Energy Relations in the Continuous-Time Fourier Series

A hard-limiter is a device whose output is the sign of the incoming signal as a function of time. Specifically, when the input signal $x(t)$ is positive, the output signal $y(t)$ is equal to 1, and when $x(t)$ is negative, $y(t)$ equals $-1$. Some implementations of frequency modulation (FM) or radar systems use a hard-limiter to process the phase of an incoming signal while ignoring any possible amplitude distortion. In this exercise, you will consider passing the signal $x(t) = \cos(\omega_0 t + \phi)$ through a hard limiter to produce $y(t)$.

### Intermediate Problems

(a). Find the CTFS representation for the signal $y(t)$. Hint: Use the properties of CTFS and knowledge that the symmetric square wave with period $T$,

$$s(t) = \begin{cases} 1, & |t| < T/4, \\ 0, & T/4 \le |t| \le T, \end{cases} \tag{3.9}$$

has CTFS coefficients $a_k$ given by

$$a_k = \frac{\sin(\pi k/2)}{\pi k}.$$

(b). The energy in the first fundamental frequency of a periodic signal can be defined as $|a_1|^2 + |a_{-1}|^2$, where $a_k$ is the CTFS of the signal. Calculate the energy in the first fundamental for both the output $y(t)$ and the input $x(t)$. Is there an energy gain or loss? Can you account for the energy change?

(c). Use Parseval's relation to find the total energy in a period of the signal. Approximate this sum using the first 100 frequencies, i.e., $a_k, |k| < 100$. To what value does this sum converge? Hint: Use Parseval's theorem to calculate a closed-form answer in the time-domain.

(d). To observe how quickly the energy estimate converges, plot the estimate of the signal energy as a function of the number of terms used in the sum. You may find the function `cumsum` helpful to create a vector of the partial sums,

$$e_n = \sum_{k=-n}^{n} |a_k|^2.$$

## Advanced Problem

(e). Use the analytic expression for the signal energy to find a closed-form expression for the sum

$$\sum_{k=0}^{\infty} \frac{1}{(2k+1)^2}.$$

Hint: Use Parseval's Theorem to relate the expressions for the total energy in the time and frequency domains.

## 3.8 First-Order Recursive Discrete-Time Filters

This exercise demonstrates the effect of first-order recursive discrete-time filters on periodic signals. You will examine the frequency responses of two different systems and also construct a periodic signal to use as input for these systems. This exercise assumes you are familiar with using `fft` and `ifft` to compute the DTFS of a periodic signal as described in Tutorial ??. In addition, it is also assumed you are proficient with the `filter` and `freqz` commands described in Tutorials ?? and ??. Several parts of this exercise require you to generate vectors which should be purely real, but have very small imaginary parts due to roundoff errors. Use `real` to remove these residual imaginary parts from these vectors.
This exercise focuses on two causal LTI systems described by first-order recursive difference equations:

$$\begin{aligned} \text{System 1:} \quad y[n] - 0.8y[n-1] &= x[n], \\ \text{System 2:} \quad y[n] + 0.8y[n-1] &= x[n]. \end{aligned}$$

The input signal $x[n]$ will be the periodic signal with period $N = 20$ described by the DTFS coefficients

$$a_k = \begin{cases} 3/4\,, & k = \pm 1\,, \\ -1/2\,, & k = \pm 9\,, \\ 0\,, & \text{otherwise}\,. \end{cases} \tag{3.10}$$

### Intermediate Problems

(a). Define vectors `a1` and `b1` for the difference equation describing System 1 in the format specified by `filter` and `freqz`. Similarly, define `a2` and `b2` to describe System 2.

(b). Use `freqz` to evaluate the frequency responses of Systems 1 and 2 at 1024 points between 0 and $2\pi$. Note that you will have to use the `'whole'` option with `freqz` to do this. Use `plot` and `abs` to generate appropriately labeled graphs of the magnitude of the frequency response for both systems. Based on the frequency response plots, specify whether each system is a highpass, lowpass, or bandpass filter.

(c). Use Eq. (??) to define the vector `a_x` to be the DTFS coefficients of $x[n]$ for $0 \leq k \leq 19$. Generate a plot of the DTFS coefficients using `stem` where the x-axis is labeled with frequency $\omega_k = (2\pi/20)k$. Compare this plot with the frequency responses you generated in Part ??, and for each system state which frequency components will be amplified and which will be attenuated when $x[n]$ is the input to the system.

(d). Define `x_20` to be one period of $x[n]$ for $0 \leq n \leq 19$ using `ifft` with `a_x` as specified in Tutorial ??. Use `x_20` to create `x`, consisting of 6 periods of $x[n]$ for $-20 \leq n \leq 99$. Also define `n` to be this range of sample indices, and generate a plot of $x[n]$ on this interval using `stem`.

(e). Use `filter` to compute `y1` and `y2`, the outputs of Systems 1 and 2 when $x[n]$ is the input. Plot both outputs for $0 \leq n \leq 99$ using `stem`. Based on these plots, state which output contains more high frequency energy and which has more low frequency energy. Do the plots confirm your answers in Part ???

(f). Define `y1_20` and `y2_20` to be the segments of `y1` and `y2` corresponding to $y_1[n]$ and $y_2[n]$ for $0 \leq n \leq 19$. Use these vectors and `fft` to compute `a_y1` and `a_y2`, the DTFS coefficients of `y1` and `y2`. Use `stem` and `abs` to generate plots of the magnitudes of the DTFS coefficients for both sequences. Do these plots agree with your answers in Part ???

## 3.9 Frequency Response of a Continuous-Time System

This exercise demonstrates the effect of the frequency response of a continuous-time system on periodic signals. You will examine the response of a simple linear system to each of the harmonics that compose a periodic signal as well as to the periodic signal itself. In this exercise you will need to use the function `lsim` as discussed in the Tutorial ??.

Consider a simple RC circuit that has a system function given by

$$H(s) = \frac{1}{1 + RCs},$$

whose input is given by

$$x(t) = \cos(t)$$

and whose output is $y(t)$. For the problems that follow, use `t=linspace(0,20,1000)` for all simulations, and assume that the time constant $RC$ is 1.

## Intermediate Problems

(a). Use the function `lsim` to simulate the response of the system $H(s)$ to $x(t)$ over $0 \leq t \leq 20$, storing the response in the signal $y(t)$. Plot the output $y(t)$ and the input $x(t)$ for $10 \leq t \leq 20$ on the same graph, and note the amplitude and phase change from input to output. Can you explain each of these effects in terms of the frequency response? Hint: Use the system function $H(s)$ to determine the exact form of the output $y(t)$ when the input is $x(t) = \cos(t)$.

(b). Now look at the response of the system to the square wave that results from first passing $x(t)$ through a hard-limiter, $x_2(t) = \text{sign}(\cos(t))$. A simple way to use selective indexing to generate the square wave is

```
>> x2=cos(t);
>> x2(x2>0)=ones(size(x2(x2>0)));
>> x2(x2<0)=-ones(size(x2(x2<0)));
```

Create the signal `x2` and use `lsim` to simulate the response of the circuit to the square wave. Plot the resulting output $y_2(t)$ over the interval, $10 \leq t \leq 20$.

## Advanced Problems

(c). Analytically calculate the CTFS for the square wave `x2`. You may find it helpful to first find a relationship between the signal $x_2(t)$ and the signal $s(t)$ defined in Eq. (??). Use the ten lowest frequency nonzero CTFS coefficients of `x2` to create the first 5 harmonic components individually. For example, if you have the positive and negative CTFS coefficients stored in the vectors `apos_k` and `aneg_k`, respectively, you could construct the first harmonic of the input as follows:

```
>> s1=apos_k(1)*exp(j*t)+aneg_k(1)*exp(-j*t);
```

Construct the signals `s1`, `s2`, `s3`, `s4`, and `s5`. Plot the sum of these signals on the same graph as the square wave `x2`.

(d). Since the circuit is linear, the response of the system to the square wave input can be calculated by finding the response of the system to each harmonic component separately, and then summing the results. Verify this by using `lsim` to find the responses y1, ..., y5 to the signals s1, ..., s5 as well as the response of the system to the signal ssum which is the sum of the first five harmonic components, s1 through s5.

(e). Compare the response of the system to the sum of the first 5 harmonics to the response of the system to the original square wave. Can you explain why the two responses are so similar? Hint: Consider the energy in the CTFS of x2 as a function of the number of coefficients used in the approximation to x2.

(f). Verify that your signals y1, ..., y5 are correct by constructing each signal from the system function $H(s)$ and the CTFS for x2. For each, plot both the analytically determined and the simulated signals over $10 \leq t \leq 20$.

## 3.10 Computing the Discrete-Time Fourier Series

### The FFT Algorithm for Computing the DTFS

The DTFS for a periodic discrete-time signal with fundamental period $N$ is given by Eq. (??).

### Advanced Problems

(a). Argue that computing each coefficient $a_k$ requires $N+1$ complex multiplications and $N-1$ complex additions. Assume that the functions $e^{-jk(2\pi/N)n}$ do not require any computation. Using this result, how many operations are required to compute the DTFS for a signal with fundamental period $N$? Note that the number of operations required is independent of the particular signal $x[n]$.

While the DTFS is useful for analyzing many signals and systems, its popularity is due in part to the existence of a fast algorithm known as the Fast Fourier Transform (FFT). Before the landmark paper by Cooley and Tukey[1] was published in 1965, the fastest algorithms used at the time for computing the DTFS coefficients required $\mathcal{O}(N^2)$ operations. ($\mathcal{O}(N^2)$ operations means that the number of complex additions and multiplications required is a polynomial in $N$ of order 2. When characterizing the growth in computational complexity of an algorithm, the lower order terms can be ignored, since they become negligible for large $N$.) The FFT algorithm proposed by Cooley and Tukey, however, requires only $\mathcal{O}(N \log N)$ operations. For large $N$, the computational savings can be tremendous; e.g., for $N = 800$, compare $N^2 = 1 \times 10^6$ with $N \log N = 3 \times 10^3$. The difference is three orders of magnitude. The FFT essentially allows Fourier analysis to be applied to signals with very large fundamental periods. In the following problem, you will plot the number of operations required by the FFT and compare them with an $\mathcal{O}(N^2)$ algorithm.
For the next two parts, you should assume $x[n]$ has fundamental period $N$ and takes on the values $x[n] = (0.9)^n$ on the interval $0 \leq n \leq N-1$.

(b). If you have not already done the Advanced Problem in Exercise ?? writing dtfs, do so now. You will compare the amount of computation this algorithm requires with

[1] J.W. Cooley and J.W. Tukey, "An algorithm for the machine calculation of complex Fourier series", Mathematics of Computation, 1965, 19(89-92), pp. 297-301.

the amount required by `fft`. While the number of operations performed cannot be explicitly counted in MATLAB 6, the command `etime` will allow you to measure the elapsed time between the start and finish of your implementation of Eq. (??) as follows:

```
>> x = (0.9).^[0:N-1]; % create one period of x[n]
>> t0=clock;  % set t0 to the current time
>> X = dtfs(x,0); % Store the DTFS of x[n] in X
>> c = etime(clock,t0); % Store the elapsed time in c
```

Find `c` for computing `X` using `dtfs` for $N = 8$, 32, 64, 128, and 256. Save these values in the vector `dtfstime`.

(c). Now, compute the DTFS coefficients of $x[n]$ for $N = 8$, 32, 64, 128, and 256 using `fft` as shown in Tutorial ??. Use `etime` to find the elapsed time for each value of $N$ and store these values in the vector `ffttime`. Plot `dtfstime` and `ffttime` versus $N$ using `loglog`. How does the elapsed time required by `fft` compare to that required by `dtfs`, particularly for large values of N?

If computational savings is defined as the ratio of the operations required by the slow algorithm to those of the fast algorithm, note that the savings provided by the FFT increase as $N$ increases (at the rate of roughly $N/\log N$).

## Periodic Convolution with the FFT

In many applications, LTI systems are implemented using periodic convolution. In the following problems, you will implement the periodic convolution of two discrete-time signals using two different methods, one of which takes advantage of the computational savings provided by the FFT. The periodic convolution of two periodic discrete-time signals $x[n]$ and $h[n]$, both with fundamental period $N$, is

$$y[n] = \sum_{r=0}^{N-1} x[r]h[n-r]\,. \tag{3.11}$$

(d). What is the fundamental period, $N_y$, of $y[n]$? Argue that directly implementing the periodic convolution according to Eq. (??) requires $\mathcal{O}(N^2)$ operations (additions and multiplications). Remember that computing one period of $y[n]$ is sufficient to characterize the entire signal.

(e). Assume both $x[n]$ and $h[n]$ have fundamental period $N = 40$, and are given by $x[n] = (0.9)^n$ and $h[n] = (0.5)^n$ over the interval $0 \leq n \leq N-1$. Compute the periodic convolution of $x[n]$ with $h[n]$ and plot $y[n]$ for $0 \leq n \leq N_y - 1$. Store the elapsed time, given by `etime`, required to implement the convolution in `f40c`. Hint: To implement the periodic convolution, first store $x[n]$ and $h[n]$ over the interval $0 \leq n \leq N-1$ in the row vectors `x` and `h`, respectively, and then use `conv([x x],h)`. The periodic convolution can be extracted from a portion of this signal.

(f). Repeat Part ?? for $N = 80$, again plotting a period of $y[n]$ and storing the elapsed time in `f80c`.

(g). Assume $x[n]$ and $h[n]$ are defined as in Part ??. Use the periodic convolution property of the DTFS to implement the periodic convolution[2]. Namely, compute the DTFS coefficients of both $x[n]$ and $h[n]$ using `fft` as described in Tutorial ??. Then, use the periodic convolution property to form the DTFS coefficients of $y[n]$. Finally, synthesize $y[n]$ from the DTFS coefficients using `ifft`. The `ifft` algorithm is nearly identical to the FFT, and also requires $\mathcal{O}(N \log N)$ operations for a signal with fundamental period $N$. To check the validity of your implementation, plot $y[n]$ for $0 \leq n \leq N_y - 1$ and compare this signal to that computed in Part ??. Remember, as discussed in Tutorial ??, the signal $y[n]$ might have a small imaginary component due to numerical round-off errors. Store the total elapsed time required to compute $y[n]$ in `f40f`.

(h). Repeat Part ?? for $N = 80$, again plotting a period of $y[n]$ and storing the elapsed time in `f80f`. Again check the validity of your implementation by comparing $y[n]$ with that computed in Part ??.

(i). Compute the ratios of `f40c` to `f40f` and `f80c` to `f80f`. How do these ratios compare for $N = 40$ and $N = 80$? Which method of convolution is more efficient for each value of $N$? Which method would you choose for $N > 80$? Justify your answer.

## ■ 3.11 Synthesizing Continuous-Time Signals with the Fourier Series Ⓢ

A large class of continuous-time periodic signals can be represented by the sum

$$x(t) = \sum_{k=-\infty}^{\infty} a_k e^{ik(2\pi/T)t}, \tag{3.12}$$

where $a_k$ is the continuous-time Fourier series (CTFS). Equation (??) is commonly referred to as the CTFS synthesis equation. For this exercise, you will synthesize signals which have a small number of nonzero coefficients. Exercise ?? considers the Fourier analysis and synthesis of continuous-time signals with an infinite number of nonzero CTFS coefficients.

### Basic Problems

For these problems, you will construct symbolic expressions for periodic signals with a small number of nonzero Fourier series coefficients. The fundamental period and nonzero CTFS coefficients for three signals are given by

(a). $x_1(t)$: $T = 1$; $a_1 = a_{-1} = 5$, $a_3 = a_{-3} = 2$;

(b). $x_2(t)$: $T = 2$; $a_1 = a_{-1}^* = i$, $a_2 = a_{-2}^* = -\frac{1}{2}i$, $a_3 = a_{-3}^* = \frac{1}{4}i$, $a_4 = a_{-4}^* = -\frac{1}{8}i$;

(c). $x_3(t)$: $T = 4$; $a_1 = a_{-1}^* = i$, $a_2 = a_{-2}^* = \frac{1}{2}i$, $a_3 = a_{-3}^* = \frac{1}{4}i$, $a_4 = a_{-4}^* = \frac{1}{8}i$.

[2] See, for instance, Table 3.2 in *Signals and Systems*.

For each signal, create a symbolic expression for the continuous-time signal and plot the signal over two periods using `ezplot`.
Given the plot for $x_2(t)$, how could the plot for $x_3(t)$ be predicted from the Fourier series coefficients of the two signals? How could you have predicted from the Fourier coefficients that each of the three signals would be real?

## Intermediate Problems

Define

$$\text{sign}(k) = \begin{cases} 1, & k > 0, \\ 0, & k = 0, \\ -1, & k < 0, \end{cases}$$

and consider signals which have the following fundamental period $T$ and Fourier series coefficients $\{a_k\}$:

(d). $x_4(t)$ :

$$T = 5; \qquad a_k = \begin{cases} \dfrac{1}{(k^2+1)}, & |k| \leq K, \\ 0, & |k| > K, \end{cases}$$

(e). $x_5(t)$ :

$$T = 20; \qquad a_k = \begin{cases} \text{sign}(k)\dfrac{i}{(-2)^{|k|}}, & |k| \leq K, \\ 0, & |k| > K, \end{cases}$$

(f). $x_6(t)$ :

$$T = 5; \qquad a_k = \begin{cases} \dfrac{1}{2^{|k+2|}}, & |k| \leq K, \\ 0, & |k| > K. \end{cases}$$

For each signal, let $K = 1$, 3, and 9. For each value of $K$, create a symbolic expression for $x_n(t)$, $(n = 4, 5, 6)$ and plot the signals over two periods using `ezplot`. If the signal is complex, plot the real and imaginary components separately. Hint: Before using `symsum` to create a symbolic expression for $x_i(t)$, use `symmul` to create a symbolic expression for $a_k e^{ik(2\pi/T)t}$, which will be a function of both $t$ and $k$. Also, `compose` can be used to select the real or even components of a symbolic expression, i.e., `compose('real(x)',x1)` takes the real component of the symbolic expression, `x1`, for $x_1(t)$.

(g). How could you have anticipated from the Fourier series coefficients which signals would be real?

### Advanced Problem

(h). For this part, you will write an M-file which synthesizes a signal $x(t)$ from its CTFS coefficients when the coefficients are zero outside the interval $-K \leq k \leq K$. The first line of your M-file should read

```
function x = ctsynth(a,T,K)
```

where $T$ is the fundamental period of the signal $x(t)$ and a is a symbolic array containing the CTFS coefficients $a_k$ for $-K \leq k \leq K$. Be sure that your routine checks that a is a symbolic array with $2K + 1$ elements. Your function should return a symbolic expression x for the signal $x(t)$. Verify your function on some of the signals in the Basic and Intermediate Problems. For instance, the symbolic expression for $x_1(t)$ in Part **??** should be created by

```
>> T = 1;
>> K = 3;
>> a = sym('[2 0 5 0 5 0 2]');
>> x = ctsynth(a,T,K);
```

## ■ 3.12 The Fourier Representation of Square and Triangle Waves Ⓢ

The Fourier series synthesis equation for a continuous-time signal $x(t)$ with fundamental period $T$ is given by Eq. (**??**). The Fourier series coefficients can, and generally do, have an infinite number of nonzero values. For instance, any signal with a discontinuity will have a Fourier series representation with an infinite number of nonzero coefficients, which is infeasible for numeric computations. The finite sum

$$x_N(t) = \sum_{k=-N}^{N} a_k e^{ik(2\pi/T)t} \tag{3.13}$$

is often a very good approximation for some relatively small integer $N$. Equation (**??**) is often called the truncated Fourier series representation of $x(t)$.
In this exercise, you will analyze two common continuous-time signals— the periodic square wave and the periodic triangle wave—with the Fourier series. For each signal, you will consider the truncated Fourier Series synthesis equation. In particular, you will consider how $x_N(t)$ converges as $N$ increases.

### The Square Wave

Consider a periodic square wave with fundamental period $T = 2$. Over the interval $-1 < t < 1$, the square wave is described by

$$x(t) = \begin{cases} 1, & |t| \leq \frac{1}{2}, \\ 0, & |t| > \frac{1}{2}, \end{cases}$$

as illustrated in Figure **??**. In this exercise, you will analyze the Fourier series representation of the square wave, focusing upon the neighborhood of the discontinuities in the square wave.

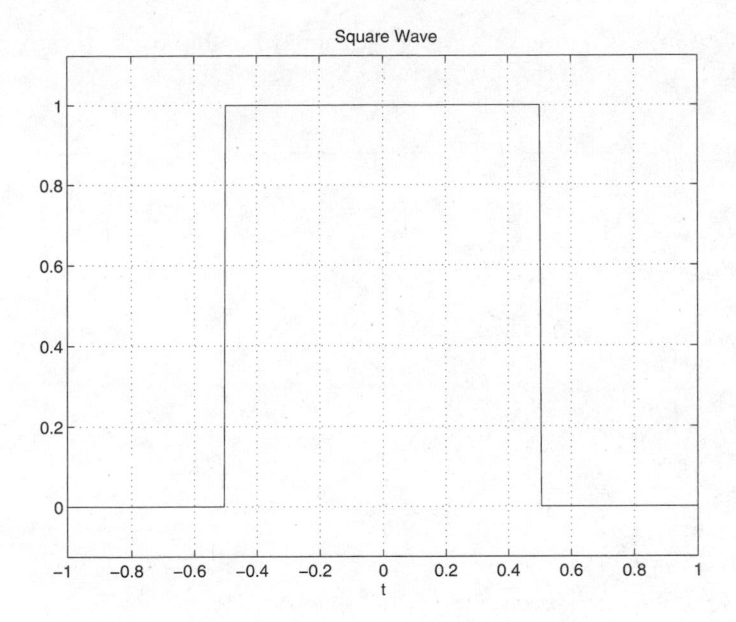


**Figure 3.6.** A single period of the square wave.

(a). Using `int`, create a symbolic expression `a` which contains the Fourier series coefficients of the square wave for each value of $k$. The symbolic expression will be a function of $k$, e.g., $a_5$ is given by `numeric(subs(a,5,'k'))`. Simplify this expression as much as possible. Use `numeric` and `stem` to plot the Fourier series coefficients for $-10 \leq k \leq 10$. Note: Depending upon how you simplify your symbolic expression for `a`, MATLAB may not be able to evaluate `a` for $k = 0$ since MATLAB cannot evaluate expressions like `sin(k)/k` at $k = 0$. In this case, calculate $a_0$ separately.

(b). Create the symbolic expression for $x_N(t)$ for $N = 1$, 3, 5, and 9. Using `ezplot`, plot $x_N(t)$ on the interval $-1 \leq t \leq 1$. Plot all four signals on the same figure using `hold`.

(c). What is the value of $x_N(t)$ at $t = \pm 1/2$? Does the value change as $N$ increases?

(d). Without explicitly evaluating $x_N(t)$, estimate the value of the overshoot error for each value of $N$. Does the overshoot error decrease as $N$ increases? How do you expect this value to change as $N \longrightarrow \infty$? (If you have the time and resources, you might plot $x_N(t)$ for $N > 9$.)

## The Triangle Wave

Consider a periodic triangle wave with fundamental period $T = 2$. Over the interval $-1 < t < 1$ the triangle wave is described by $x(t) = 1 - |t|$, as illustrated in Figure ??. While the square wave has a zeroth-order discontinuity, the triangle wave has a first-order discontinuity, i.e., a discontinuity in slope. In the following problems, you will analyze the Fourier series representation of the triangle wave and compare its behavior with the Fourier series representation of the square wave. You should be able to do these problems using simple modifications of the script file you used for Parts ??-??.

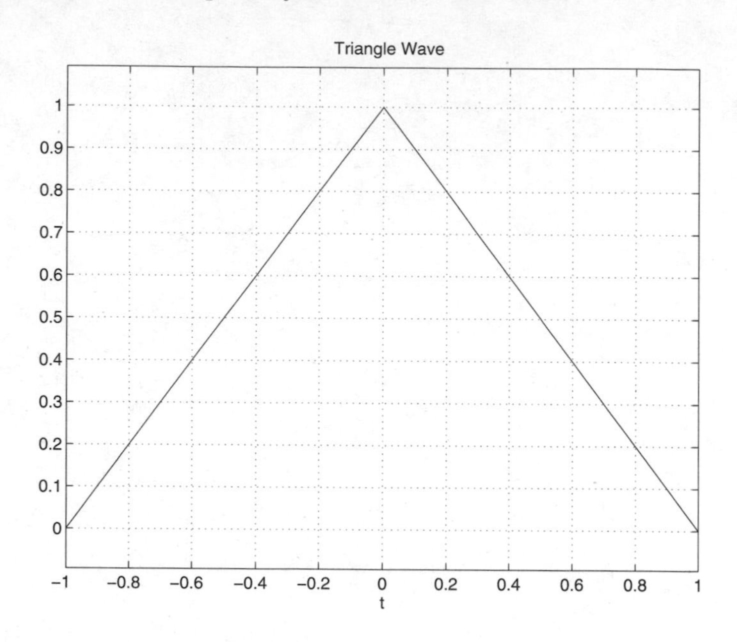


**Figure 3.7.** A single period of the triangle wave.

## Advanced Problems

(e). Using `int`, create a symbolic expression `a` which contains the Fourier series coefficients of the triangle wave for each value of $k$. The symbolic expression will be a function of $k$, e.g., $a_5$ is given by `numeric(subs(a,5,'k'))`. Simplify this expression as much as possible. Use `numeric` and `stem` to plot the Fourier series coefficients for $-10 \leq k \leq 10$. Note: MATLAB may not be able to evaluate `a` for $k = 0$ since MATLAB cannot evaluate expressions like `sin(k)/k` at $k = 0$. In this case, calculate $a_0$ separately.

(f). Create the symbolic expression for $x_N(t)$ for $N = 1$, 3, 5, and 9. Using `ezplot`, plot $x_N(t)$ on the interval $-1 \leq t \leq 1$. Use a different plot for each value of $N$.

(g). How does $x_N(t)$ appear to converge for increasing $N$ at $t = 0$? Does the maximum error

$$\max_{-1<t<1} |x(t) - x_N(t)|$$

appear to decrease as $N$ increases? How does this behavior compare with the truncated Fourier series approximation of the square wave?

(h). You can use `diff` to analyze how well $x_N(t)$ approximates the derivative of the triangle wave. For $N = 9$, use `diff` to create the symbolic expression for $dx_N(t)/dt$ from that for $x_N(t)$. Plot the derivative for $-1 < t < 1$. How does this signal compare with the $x_9(t)$ plotted in Part ??? How can you explain the similarity?

## 3.13 Continuous-Time Filtering Ⓢ

Consider the RL circuit illustrated in Figure ??. The differential equation relating the input voltage $v_s$ to the resistor voltage $v_r$ is

$$\frac{L}{R}\frac{dv_r}{dt} + v_r = v_s\,.$$

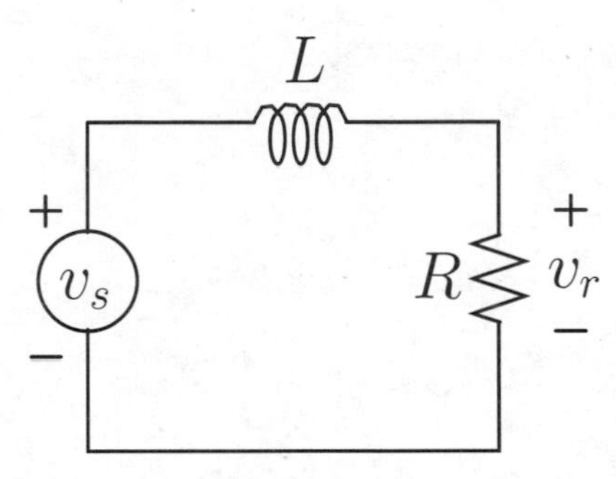


**Figure 3.8.** RL circuit with $v_s$ as the input (source).

### Basic Problems

(a). Assuming the circuit is initially at rest, the input-output relationship between $v_s$ and $v_r$ can be described by a linear time-invariant (LTI) system. Derive the corresponding frequency response $H(j\omega)$ by calculating $v_r$ in response to the complex exponential input $v_s(t) = e^{j\omega t}$.

(b). Assume that $R = 10\ \Omega$ and that $L = 1\ H$ for the rest of the exercise. Create a symbolic expression for $H(j\omega)$. On separate figures, use `ezplot` to plot the magnitude and phase of $H(j\omega)$. Be sure that a significant portion of both the passband and stopband is displayed. Is the system a highpass or lowpass filter?

(c). Without using MATLAB or making any numerical calculations, use the eigenfunction property of LTI systems to determine the resistor voltage in response to the following inputs:

(i) $v_s(t) = \cos(\pi t)$,

(ii) $v_s(t) = \sin(2t)$.

### Intermediate Problems

(d). Use `dsolve` to solve the differential equation for each of the inputs given in Part ??. Be sure to use `simple` to simplify your expressions. How do you explain the difference between the solutions predicted in Part ?? and those given by `dsolve`?

(e). Use `step` and `impulse` as described in Tutorial ?? to plot the step response and impulse response, respectively, of the resistor voltage. Explain the step response in terms of the properties of the resistor and inductor.

Chapter 4

# The Continuous-Time Fourier Transform

The continuous-time Fourier transform (CTFT)

$$x(t) = \frac{1}{2\pi}\int_{-\infty}^{\infty} X(j\omega)e^{j\omega t}d\omega \tag{4.1}$$

$$X(j\omega) = \int_{-\infty}^{\infty} x(t)e^{-j\omega t}dt \tag{4.2}$$

extends the continuous-time Fourier series (CTFS) to allow frequency-domain analysis of aperiodic as well as periodic continuous-time signals. This is an important and powerful technique since many signals that appear to have complicated structure when viewed in the time domain are simple when viewed in the frequency domain. In addition, the behavior of many LTI systems is easier to understand in the frequency domain than in the time domain. To use frequency-domain techniques effectively, it is important to develop intuition for how properties of signals in the time and frequency domains are related. The exercises in this chapter will help to foster this intuition for signals in general, and for the impulse responses and frequency responses of LTI systems in particular.
Tutorial ?? explains how `freqs` can be used to compute the frequency response of continuous-time systems. A method for computing a numerical approximation to the CTFT is shown in Exercise ??. Properties of the CTFT are demonstrated in Exercise ?? with an audio signal. Exercise ?? covers the relationship between the impulse response and frequency response of lowpass filters. The function `residue` is helpful for computing the impulse response of complicated system functions as shown in Exercise ??. Exercise ?? uses amplitude modulation to decode Morse Code messages. Finally, Exercise ?? computes the CTFT of several different signals symbolically with the function `fourier`.

## 4.1 Tutorial: `freqs`

A stable LTI system is completely characterized in terms of its frequency response, $H(j\omega)$. If $X(j\omega)$ is the CTFT of the system input, then

$$Y(j\omega) = H(j\omega)X(j\omega)$$

provides the CTFT of the system output. LTI systems whose input and output satisfy linear constant-coefficient differential equations are an important class of systems, in part because the frequency response of such systems is easily determined. Thus, for the LTI system whose input and output satisfy

$$\sum_{k=0}^{N} a_k \frac{d^k y(t)}{dt^k} = \sum_{m=0}^{M} b_m \frac{d^m x(t)}{dt^m}, \tag{4.3}$$

the frequency response follows directly as

$$H(j\omega) = \frac{b_M (j\omega)^M + b_{M-1} (j\omega)^{M-1} + \ldots + b_1 (j\omega) + b_0}{a_N (j\omega)^N + a_{N-1} (j\omega)^{N-1} + \ldots + a_1 (j\omega) + a_0}. \tag{4.4}$$

The function `freqs(b,a)` can be used to calculate and plot this frequency response, where the vectors `b` and `a` contain the coefficients $b_m$ and $a_k$, respectively. The ordering of the coefficients in `b` and `a` is exactly the same as the ordering required for the inputs to `lsim` (see Tutorials ?? and ??). This ordering is given by $\texttt{b}(M-m+1) = b_m$ and $\texttt{a}(N-k+1) = a_k$. Consider the first-order differential equation

$$\frac{dy(t)}{dt} + 3\,y(t) = 3\,x(t),$$

which describes the input-output relationship of a causal, stable LTI system. The frequency response of this system is

$$H(j\omega) = \frac{3}{3+j\omega}. \tag{4.5}$$

If no output argument is supplied to `freqs(b,a)`, the magnitude and phase of $H(j\omega)$ will be automatically plotted. Executing the commands

```
>> a = [1 3];
>> b = 3;
>> freqs(b,a)
```

produces the plot given in Figure ??. Use Eq. (??) to convince yourself that the magnitude and phase plots in Figure ?? are correct.

The command `freqs` automatically selects the range of frequencies over which $H(j\omega)$ is plotted. If you desire $H(j\omega)$ at particular values of $\omega$, or would like to plot $H(j\omega)$ over a different range of frequencies than is automatically selected by `freqs`, you can supply these frequencies as inputs. For example, executing

```
>> w = linspace(0,3*pi);
>> H = freqs(b,a,w);
```

returns $H(j\omega)$ for one hundred uniformly spaced samples of $\omega$ on the interval $0 \leq \omega \leq 3\pi$. These samples of $H(j\omega)$ are returned in `H` at the frequencies in `w`. The plot of the magnitude of `H` is given in Figure ??. See if you can replicate this plot.

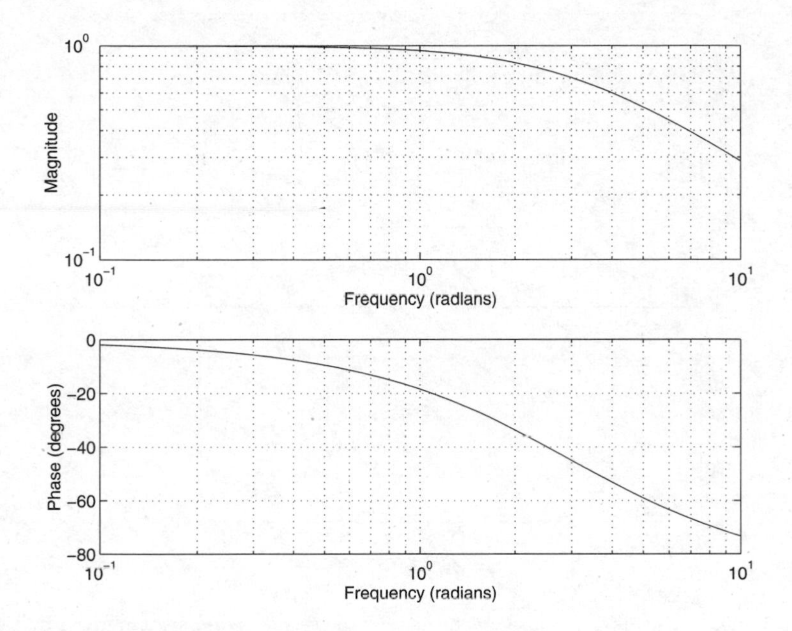


**Figure 4.1.** The frequency response as displayed by `freqs(b,a)`. Note that the frequency axes and magnitude axis are scaled logarithmically, and that only a finite interval of the frequency response is plotted.

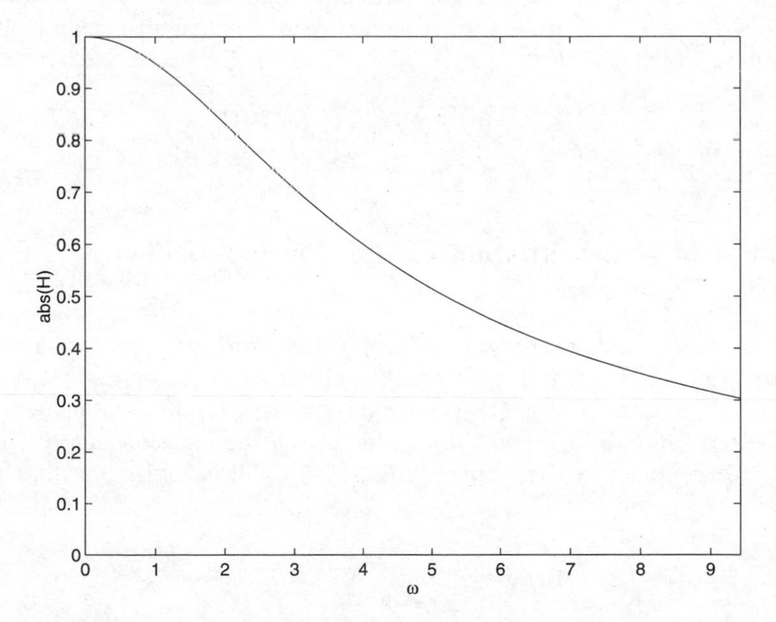


**Figure 4.2.** The frequency response magnitude $|H(j\omega)|$ for a first-order system when evaluated at the frequencies in `w=linspace(0,3*pi)`.

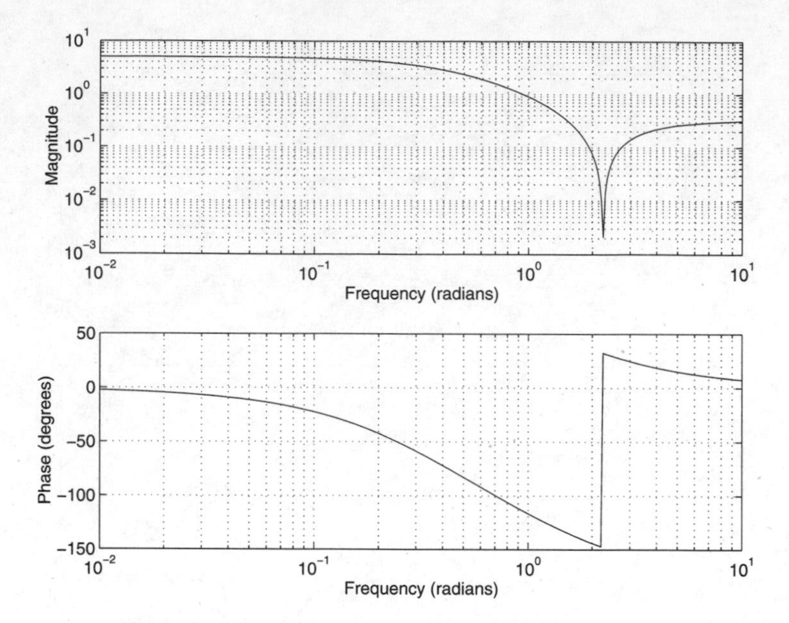


**Figure 4.3.** The frequency response magnitude for the second-order system given by Eq. (??).

Of course, `freqs` can also be used to compute or plot the frequency response of LTI systems described by differential equations of higher-order. Using `freqs`, try to reproduce the plots given in Figure ??, which contain the frequency response magnitude and phase for the stable LTI system which satisfies

$$3\,\frac{d^2y(t)}{dt^2}+4\,\frac{dy(t)}{dt}+y(t)=\frac{d^2x(t)}{dt^2}+5\,x(t)\,. \tag{4.6}$$

## 4.2 Numerical Approximation to the Continuous-Time Fourier Transform

A large class of signals can be represented using the continuous-time Fourier transform (CTFT) in Eq. (??). In this exercise you will use MATLAB to compute numerical approximations to the CTFT integral, Eq. (??). By approximating the integral using a summation over closely spaced samples in $t$, you will be able to use the function `fft` to compute your approximation very efficiently. The approximation you will use follows from the definition of the integral

$$\int_{-\infty}^{\infty} x(t)e^{-j\omega t}dt=\lim_{\tau\to 0}\sum_{n=-\infty}^{\infty} x(n\tau)e^{-j\omega n\tau}\tau.$$

For a large class of signals and for sufficiently small $\tau$, the sum on the right-hand side is a good approximation to the CTFT integral. If the signal $x(t)$ is equal to zero for $t<0$ and

$t \geq T$, then the approximation can be written

$$\int_{-\infty}^{\infty} x(t)e^{-j\omega t}dt = \int_{0}^{T} x(t)e^{-j\omega t}dt \approx \sum_{n=0}^{N-1} x(n\tau)e^{-j\omega n\tau}\tau, \tag{4.7}$$

where $T = N\tau$ and $N$ is an integer. You can use the function `fft` to compute the sum in Eq. (??) for a discrete set of frequencies $\omega_k$. If the $N$ samples $x(n\tau)$ are stored in the vector `x`, then the function call `X=tau*fft(x)` calculates

$$\mathtt{X}(k+1) = \tau \sum_{n=0}^{N-1} x(n\tau)e^{-j\omega_k n\tau}, \quad 0 \leq k < N, \tag{4.8}$$

$$\approx X(j\omega_k), \tag{4.9}$$

where

$$\omega_k = \begin{cases} \dfrac{2\pi k}{N\tau}, & 0 \leq k \leq \dfrac{N}{2}, \\ \dfrac{2\pi k}{N\tau} - \dfrac{2\pi}{\tau}, & \dfrac{N}{2}+1 \leq k < N, \end{cases} \tag{4.10}$$

and $N$ is assumed to be even. For reasons of computational efficiency, `fft` returns the positive frequency samples before the negative frequency samples. To place the frequency samples in ascending order, you can use the function `fftshift`. To store in `X` samples of $X(j\omega_k)$ ordered such that $\mathtt{X}(k+1)$ is the CTFT evaluated at $-\pi/\tau + (2\pi k/N\tau)$ for $0 \leq k \leq N-1$, use `X=fftshift(tau*fft(x))`.
For this exercise, you will approximate the CTFT of $x(t) = e^{-2|t|}$ using the function `fft` and a truncated version of $x(t)$. You will see that for sufficiently small $\tau$, you can compute an accurate numerical approximation to $X(j\omega)$.

## Basic Problems

(a). Find an analytic expression for the CTFT of $x(t) = e^{-2|t|}$. You may find it helpful to think of $x(t) = g(t) + g(-t)$, where $g(t) = e^{-2t}u(t)$.

(b). Create a vector containing samples of the signal $y(t) = x(t-5)$ for $\tau = 0.01$ and $T = 10$ over the range `t=[0:tau:T-tau]`. Since $x(t)$ is effectively zero for $|t| > 5$, you can calculate the CTFT of the signal $y(t) = x(t-5)$ from the above analysis using $N = T/\tau$. Your vector `y` should have length $N$.

(c). Calculate samples $Y(j\omega_k)$ by typing `Y=fftshift(tau*fft(y))`.

(d). Construct a vector `w` of frequency samples that correspond to the values stored in the vector `Y` as follows:

```
>> w = -(pi/tau)+(0:N-1)*(2*pi/(N*tau));
```

(e). Since $y(t)$ is related to $x(t)$ through a time shift, the CTFT $X(j\omega)$ is related to $Y(j\omega)$ by a linear phase term of the form $e^{j5\omega}$. Using the frequency vector `w`, compute samples of $X(j\omega)$ directly from `Y`, storing the result in the vector `X`.

(f). Using `abs` and `angle`, plot the magnitude and phase of `X` over the frequency range specified in `w`. For the same values of $\omega$, also plot the magnitude and phase of the analytic expression you derived for $X(j\omega)$ in Part ??. Does your approximation of the CTFT match what you derived analytically? If you plot the magnitude on a logarithmic scale, using `semilogy`, you will notice that at higher frequencies the approximation is not as good as at lower frequencies. Since you have approximated $x(t)$ with samples $x(n\tau)$, your approximation will be better for frequency components of the signal that do not vary much over time intervals of length $\tau$.

(g). Plot the magnitude and phase of `Y` using `abs` and `angle`. How do they compare with `X`? Could you have anticipated this result?

## 4.3 Properties of the Continuous-Time Fourier Transform

In this exercise, you will enhance your understanding of the continuous-time Fourier transform (CTFT) by analyzing and manipulating audio signals in the frequency and time domains. Audio signals in MATLAB are represented by vectors containing samples of the continuous-time audio signal. The sampling rate is assumed to be 8192 Hz, i.e., the audio signal is sampled every $\Delta t = (1/8192)$ seconds. To be more precise, for an audio signal $y(t)$ sampled at 8192 Hz over the interval $0 \leq t < N\,\Delta t$, the $N$-element vector `y` which represents the audio signal is given by

$$\mathtt{y}(n) = y((n-1)\,\Delta t)\,, \qquad n = 1, \ldots, N\,.$$

The function `sound` can then be used to play the signal on your computer's speaker. While this is a sampled representation of a continuous-time audio signal $y(t)$, if $y(t)$ is zero outside the sampling interval and the sampling rate $f_s = 8192$ Hz is fast enough[1], then `y` can be considered an accurate representation of $y(t)$. For all the signals in this exercise, `y` can be assumed to be an accurate representation of $y(t)$. To begin this exercise, you must first load a sample audio signal by typing

```
>> load splat
>> y = y(1:8192));
```

To verify that you have accurately loaded the audio data, and to convince yourself that the MATLAB vector `y` can accurately represent an audio signal, type

```
>> N = 8192;
>> fs = 8192;
>> sound(y,fs)
```

The function `fft` takes the sampled representation `y` and computes an approximation to the CTFT of $y(t)$, $Y(j\omega)$, at samples of $\omega$. If you type

[1]In Chapter ??, you will explore how fast a signal must be sampled so that the sampled version of the signal accurately represents the underlying continuous-time signal.

```
>> Y = fftshift(fft(y));
```

then the vector Y contains an approximation to $Y(j\omega)$ at N evenly spaced frequencies on the interval $-\pi f_s \le \omega < \pi f_s$. In fact, Y contains only approximate values of $cY(j\omega)$, where $c$ is a constant, but you should not worry about the approximation or the scaling for the purposes of this exercise. Refer to Exercise **??** for a more complete discussion of the relationship between $Y(j\omega)$ and Y.
The function `fftshift` re-orders the output of `fft` so that the samples of $Y(j\omega)$ are ordered in Y from the most negative to most positive frequencies. Most of the properties that you associate with the CTFT can now be verified on the vector Y.

## Basic Problems

(a). Type `Y=fftshift(fft(y))` to calculate the Fourier transform vector Y. The corresponding frequency values can be stored in the vector w by typing

```
>> w = [-pi:2*pi/N:pi-pi/N]*fs;
```

Use w and Y to plot the magnitude of the continuous-time Fourier transform over the interval $-\pi f_s \le \omega < \pi f_s$.

The function `ifft` is the inverse operation of `fft`. For even length vectors, `fftshift` is its own inverse. For the vector Y, $N = 8192$ and the inverse Fourier transform can be found by typing

```
>> y = ifft(fftshift(Y));
>> y = real(y);
```

The `real` function is applied here because the original time-domain signal was known to be real. However, numerical roundoff errors in `fft` and `ifft` can introduce a very small nonzero imaginary component to y. In general, the inverse CTFT is not necessarily a real signal, and the imaginary part may contain significant energy. The `real` function should only be used on the output of `ifft` when you know the resulting signal must be real, e.g., an audio signal, and you have verified the imaginary part that you will be removing is insignificant.

(b). Set `Y1=conj(Y)`, and store in `y1` the inverse Fourier transform of `Y1`. Use `real(y1)` to ensure `y1` is real. Play `y1` using `sound(y1,fs)`. Can you explain what you just heard, knowing how the inverse Fourier transform of $Y^*(j\omega)$ is related to $y(t)$?

## Intermediate Problems

The CTFT of $y(t)$ can be written in terms of its magnitude and phase as

$$Y(j\omega) = |Y(j\omega)|\, e^{j\phi(\omega)},$$

where $\phi(\omega) = \angle Y(j\omega)$. For many signals, either the phase or the magnitude alone can be used to construct a useful approximation of the signal $y(t)$. For instance, consider the signals $y_2(t)$ and $y_3(t)$ with CTFTs

$$Y_2(j\omega) = |Y(j\omega)| \quad \text{and} \quad Y_3(j\omega) = e^{j\phi(\omega)}.$$

(c). Demonstrate analytically that $y_2(t)$ and $y_3(t)$ will be real signals whenever $y(t)$ is real.

(d). Construct a vector `Y2` equal to the magnitude of `Y`. Store in `y2` the inverse Fourier transform of `Y2`. Play this vector using `sound`.

(e). Construct a vector `Y3` which has the same phase as `Y`, but has magnitude equal to one for each frequency. Store in `y3` the inverse Fourier transform of `Y3`. Play this vector using `sound`.

(f). Based upon the two signals you just heard, which component of the Fourier transform would you say is most crucial for representing an audio signal: the magnitude or the phase?

## Advanced Problems

These problems consider the effect of scaling the time-axis on the CTFT. Namely, you will examine how the transformation $y(at)$ for $a > 0$ affects the Fourier transform of the signal. For $a > 1$, $y(at)$ corresponds to a compression of the time axis, while $0 < a < 1$ corresponds to an expansion of the time axis. In addition, you will see that if $y(t)$ is sampled sufficiently fast, you can process `y` to obtain the samples you would have obtained by sampling $y(at)$ for $a > 1$. You need not have the continuous-time signal $y(at)$ available to you to do this. You will also see that for $0 < a < 1$, you can process `y` in discrete-time to approximate the samples you would have obtained from $y(at)$.
If $a > 1$ and $y(t)$ is defined on the infinite interval $t \in [-\infty, \infty]$, then $y(at)$ is well defined. However, the vector `y` contains samples of $y(t)$ only over the interval $0 \leq t < N\,\Delta t$. To derive a vector `ya` from `y` which corresponds to samples of $y(at)$ with $a > 1$, the following two assumptions are made: (i) $y(t)$ is zero outside the interval $0 \leq t < N\,\Delta t$, and (ii) $a$ is an integer. The second assumption ensures that every $a$-th sample of `y` will be in `ya`.

(g). Use `y` to create a vector `y4` which contains the values you would obtain from sampling $y_4(t) = y(2t)$ at 8192 Hz. Note that sampling $y_4(t)$ at 8192 Hz corresponds to sampling $y(t)$ with a new $\Delta t_2$ equal to twice the old $\Delta t$, i.e., $\Delta t_2 = 2/8192$. The vector `y4` should contain $N$ samples, the second $N/2$ of which are equal to zero.

(h). Play `y4` using `sound(y4,fs)`. Can you explain the change in pitch by comparing the Fourier transform of `y4` with that of `y`? How does the compression of a signal affect its Fourier transform?

When $a$ was an integer greater than one, you were able to extract the desired samples of $y(at)$ directly from `y`. When $0 < a < 1$, a different problem arises. You must interpolate samples to fill in new values between the samples in `y`. To see the difficulty in interpolating these values, consider the following scenario with $a = 1/2$. Let `y5` contain the values you would have obtained by sampling $y_5(t) = y(t/2)$ at 8192 Hz on the interval $0 \leq t < N\,\Delta t$. Note that these are the same samples you would get by sampling $y(t)$ twice as fast, i.e., at 16384

Hz. Since `y5` will be twice as long as `y`, you must interpolate the additional samples of `y5` that fall between the samples in `y`. Specifically, you will use the linear interpolator discussed in Exercise ??. This will allow you to approximate the desired samples of $y_5(t) = y(t/2)$.

(i). Create the vector `x` given by

$$\texttt{x(n)} = \begin{cases} \texttt{y(n/2)}, & \text{n even}, \\ 0, & \text{n odd}. \end{cases}$$

(j). Use the function `filter` described in Tutorial ?? to perform linear interpolation on `x`. The impulse response for the linear interpolator you want to use here is `h=[1 2 1]/2`. Let `y5` be the result of the interpolation.

(k). Play `y5` using `sound(y5,fs)`. Can you explain the change in pitch by comparing the Fourier transform of `y5` with that of `y`? Remember that the Fourier transform of `y5` contains `2*N` samples of $Y_5(j\omega)$ over the interval $-\pi f_s \leq \omega < \pi f_s$. You will need to define a new frequency vector analogous to the one given in Part ?? but with $2N$ points. How does an expansion of $y(t)$ affect the Fourier transform?

## ■ 4.4 Time- and Frequency-Domain Characterizations of Systems

The impulse response $h(t)$ of an LTI system completely characterizes the system since the response $y(t)$ to any input $x(t)$ is given by the convolution $y(t) = h(t) * x(t)$. If the system is stable, an equivalent representation of the system is given by its frequency response $H(j\omega)$. In this case the continuous-time Fourier transforms are related by $Y(j\omega) = H(j\omega)X(j\omega)$. In this exercise, you will consider a number of stable LTI systems described by linear constant-coefficient differential equations. For these systems, you will be asked to calculate their impulse and frequency responses. Although either the frequency response or the impulse response is sufficient to completely characterize an LTI system, you will learn that it is sometimes advantageous to consider system properties in both the time and frequency domains.

### Basic Problems

These problems assume that you have completed Tutorial ??, on `freqs`. Consider the class of causal LTI systems given by

$$\frac{dy(t)}{dt} + a_0\, y(t) = a_0\, x(t)\,, \tag{4.11}$$

where $a_0 > 0$ guarantees stability. Define System I to be the system satisfying Eq. (??) for $a_0 = 3$ and define System II to be the system for $a_0 = 1/3$.

(a). Analytically derive the frequency response for the stable LTI system corresponding to Eq. (??). Also determine the magnitude and phase of this frequency response.

(b). Define `w=linspace(0,10)`. Use `freqs` to calculate the frequency response of Systems I and II at the frequencies in `w`. Plot the magnitude of these two responses in a single figure. Do the magnitude plots agree with your analytic expression for the frequency response magnitude?

(c). Use the function `impulse` described in Tutorial ?? to calculate the impulse response of Systems I and II at time samples defined by the vector `t=linspace(0,5)`.

(d). What is the relationship between the rate at which the impulse response decays (with time) and the rate at which the frequency response magnitude decays (with frequency)? What CTFT property explains this relationship?

## Intermediate Problems

As should be evident from the frequency responses you plotted in Part ??, the LTI systems given by Eq. (??) attenuate high frequency components more than low frequency components. In this sense, Systems I and II are lowpass filters. For some applications, one desires a sharper transition in the frequency response, i.e., a sharper transition from those frequencies which are to be passed by the filter to those which are to be attenuated. The sharpest transition is obtained by the ideal lowpass filter, which is defined by

$$H(j\omega) = \begin{cases} 1, & |\omega| \leq \omega_c, \\ 0, & |\omega| > \omega_c, \end{cases}$$

where $\omega_c$ is the cutoff frequency. The impulse response of the ideal lowpass filter reveals that this filter is noncausal:

$$h(t) = \frac{\sin(\omega_c t)}{\pi t}.$$

Real-time signal processing applications require causal filters and thus cannot use the ideal lowpass filter. However, causal filters which approximate the magnitude response of the ideal lowpass filter can be realized with linear constant-coefficient differential equations. In the following problems, you will consider the properties of one such filter.

(e). Butterworth filters[2] are one class of continuous-time frequency-selective filters which can be realized by linear constant-coefficient differential equations. To determine the coefficients of a second-order Butterworth filter with $\omega_c = 3$, type

```
>> wc = 3;
>> [b2,a2] = butter(2,wc,'s');
```

The argument `'s'`, specifies that you desire a continuous-time filter rather than the discrete-time filter that `butter` returns by default. The vectors `b2` and `a2` contain the coefficients for a second-order differential equation specified in the same format used in Tutorial ??. Use `freqs` to compute and plot the frequency response magnitude of the corresponding LTI system at frequencies `w=linspace(0,10)`. On the same

[2] Butterworth filters are described in Section 9.7.5 of *Signals and Systems.* A more detailed discussion of filter classes and filtering theory is given in *Introduction to Filter Theory* by D. E. Johnson, Prentice-Hall, 1976.

plot, include the frequency response magnitude of System I, which you computed in Part ??.

(f). In terms of frequency response magnitude, which system more closely approximates an ideal lowpass filter with $\omega_c = 3$, System I or the second-order Butterworth filter? How does the phase of the frequency response of each system compare with that of the ideal lowpass filter?

### Advanced Problems

(g). Use `impulse` to compute the impulse response of the second-order Butterworth filter for the time samples `t=linspace(0,5)`. Plot the impulse responses of System I and the second-order Butterworth filter on the same set of axes.

(h). In some practical filtering systems, you would like the filter output to be non-negative for any non-negative input, i.e., $y(t) \geq 0$ for all $-\infty < t < \infty$ when $x(t) \geq 0$ for all $-\infty < t < \infty$. Using the impulse responses plotted in Part ??, argue whether or not the output of each system can have negative values when the input is non-negative for all $-\infty < t < \infty$. If the system output can be negative, provide an example input and the corresponding output computed by `lsim`.

## 4.5 Impulse Responses of Differential Equations by Partial Fraction Expansion

In this exercise, you will learn how to find analytic expressions for the impulse responses of stable LTI systems whose inputs and outputs satisfy linear constant-coefficient differential equations. The frequency response of systems of this form can be written as ratios of polynomials in $(j\omega)$. MATLAB represents these polynomials as a vector of coefficients of the polynomial in decreasing powers of the dependent variable, $j\omega$. For example, the polynomial $G(j\omega) = 4(j\omega)^3 - 5(j\omega)^2 + 2(j\omega) - 7$ would be represented in MATLAB by the vector `G=[4 -5 2 -7]`. MATLAB contains several functions for manipulating polynomials in this format. One very useful function is `residue`, which computes the partial fraction expansion of a function consisting of a ratio of polynomials. In this exercise, you will learn to convert the differential equation relating the input and output of a stable, continuous-time, LTI system into vectors representing the polynomials appearing in the numerator and denominator of the frequency response. Then, you will use `residue` to process the frequency response so that the impulse response may be easily determined from the partial fraction expansion.

### Basic Problems

In the next several parts, you will be working with the causal, continuous-time LTI system whose input and output are related by the differential equation

$$\frac{d^2y(t)}{dt^2} + \frac{3}{2}\frac{dy(t)}{dt} + \frac{1}{2}y(t) = \frac{dx(t)}{dt} - 2x(t). \tag{4.12}$$

(a). Find $H_1(j\omega)$, the frequency response of the causal, LTI system whose input and output satisfy Eq. (??). Define vectors `b1` and `a1` to represent the numerator and denominator polynomials in $j\omega$.

(b). The command `[r1,p1]=residue(b1,a1)` will compute the partial fraction expansion of $H_1(j\omega)$ so long as the ratio of the polynomials represented by `b1` and `a1` is a proper fraction. Recall a proper fraction is one whose numerator polynomial has an order that is strictly less than the denominator polynomial order. The vector `r1` contains the numerators of the partial fraction terms and the vector `p1` contains the roots of the denominators. If the denominator does not contain any repeated roots and the system function was a proper fraction, the vectors returned by `residue` represent a sum of the form

$$\sum_{m=1}^{N} \frac{\texttt{r1}(m)}{j\omega - \texttt{p1}(m)},$$

where $N$ is the length of the vectors `r1` and `p1`. Write out the partial fraction expansion of $H_1(j\omega)$ based on the output of `residue` and recombine the terms analytically to verify they give you $H_1(j\omega)$.

(c). The impulse response $h_1(t)$ of the system is the inverse CTFT of the frequency response $H_1(j\omega)$. The partial fraction expansion makes it easy to invert each term by inspection and combine the results using linearity to obtain the impulse response. Write out the impulse response $h_1(t)$ for this system. As a check, recall that the CTFT only exists for continuous-time signals which are absolutely integrable. Is the $h_1(t)$ you obtained absolutely integrable?

## Intermediate Problems

The `residue` function is also capable of computing the partial fraction expansion for frequency responses where one of the roots of the denominator polynomial is repeated. Consider the causal LTI system whose input and output satisfy the differential equation

$$\frac{d^3y(t)}{dt^3} + 7\frac{d^2y(t)}{dt^2} + 16\frac{dy(t)}{dt} + 12y(t) = 3\frac{d^2x(t)}{dt^2} + 10\frac{dx(t)}{dt} + 5x(t).$$

(d). Define vectors `b2` and `a2` to represent the numerator and denominator polynomials of the frequency response $H_2(j\omega)$ for this system.

(e). Use `residue` to obtain vectors `r2` and `p2`. Notice that one of the elements of `p2` is repeated. This means the denominator had a second-order root at that value. The first element of `r2` corresponding to the repeated value in `p2` is the numerator for the term with denominator $(j\omega - \texttt{p2}(k))$, where $k$ is the index of the first of the repeated elements of `p2`. The next element of `r2`, $\texttt{r2}(k+1)$, is the numerator of the term with denominator $(j\omega - \texttt{p2}(k+1))^2$. Typing `help residue` explains this in more detail. Find the partial fraction expansion for $H_2(j\omega)$ and analytically recombine the terms of the sum to verify you get back the same $H_2(j\omega)$.

(f). Use the partial fraction expansion to find the impulse response $h_2(t)$. You will have to find the inverse CTFT of a term of the form

$$\frac{c}{(j\omega - a)^2}$$

in order to find $h_2(t)$. Is the impulse response $h_2(t)$ absolutely integrable?

## Advanced Problems

Consider the stable, continuous-time system whose inputs and outputs satisfy the differential equation

$$\frac{d^2y(t)}{dt^2} - 4y(t) = -4x(t).$$

(g). Define vectors `b3` and `a3` to represent the numerator and denominator polynomials of the system function $H_3(j\omega)$.

(h). Compute the partial fraction expansion of $H_3(j\omega)$ using `residue`. Analytically recombine the terms of your sum to verify you get back $H_3(j\omega)$.

(i). Determine the impulse response $h_3(t)$ for the system based on the partial fraction expansion. Remember that $h_3(t)$ must be absolutely integrable because you have assumed the system is stable. Is $h_3(t)$ causal?

# 4.6 Amplitude Modulation and the Continuous-Time Fourier Transform

This exercise will explore amplitude modulation of Morse code messages. A simple amplitude modulation system can be described by

$$x(t) = m(t)\cos(2\pi f_0 t), \tag{4.13}$$

where $m(t)$ is called the message waveform and $f_0$ is the modulation frequency. The continuous-time Fourier transform (CTFT) of a cosine of frequency $f_0$ is

$$C(j\omega) = \pi\delta(\omega - 2\pi f_0) + \pi\delta(\omega + 2\pi f_0), \tag{4.14}$$

which can confirmed by substituting $C(j\omega)$ into Eq. (??) to yield

$$\cos(2\pi f_0 t) = \frac{1}{2}\left(e^{j2\pi f_0 t} + e^{-j2\pi f_0 t}\right). \tag{4.15}$$

Using $C(j\omega)$ and the multiplication property of the CTFT, you can obtain the CTFT of $x(t)$, namely,

$$X(j\omega) = \frac{1}{2}M\big(j(\omega - 2\pi f_0)\big) + \frac{1}{2}M\big(j(\omega + 2\pi f_0)\big), \tag{4.16}$$

where $M(j\omega)$ is the CTFT of $m(t)$. Since the CTFT of a sinusoid can be expressed in terms of impulses in the frequency domain, multiplying the signal $m(t)$ by a cosine places copies of $M(j\omega)$ at the modulation frequency.
The remainder of this exercise will involve the signal,

$$x(t) = m_1(t)\cos(2\pi f_1 t) + m_2(t)\sin(2\pi f_2 t) + m_3(t)\sin(2\pi f_1 t), \tag{4.17}$$

and several parameters that can be loaded into MATLAB from the file `ctftmod.mat`. This file is in the Computer Explorations Toolbox, which can be obtained from The MathWorks at the address listed in the Preface. If the file is in one of the directories in your MATLABPATH, type `load ctftmod.mat` to load the required data. The directories contained in your MATLABPATH can be listed by typing `path`. If the file has been successfully loaded, then typing `who` should produce the following result:

```
>> who

Your variables are:

af          dash        f1          t
bf          dot         f2          x
```

In addition to the signal $x(t)$, you also have loaded

- a lowpass filter, whose frequency response can be plotted by `freqs(bf,af)`,
- modulation frequencies `f1` and `f2`,
- two prototype signals `dot` and `dash`,
- a sequence of time samples `t`.

To make this exercise interesting, the signal $x(t)$ contains a simple message. When loading the file, you should have noticed that you have been transformed into Agent 008, the code-breaking sleuth. The last words of the aging Agent 007 were "The future of technology lies in ... " at which point Agent 007 produced a floppy disk and keeled over. The floppy disk contained the MATLAB file `ctftmod.mat`. Your job is to decipher the message encoded in $x(t)$ and complete Agent 007's prediction.
Here is what is known. The signal $x(t)$ is of the form of Eq. (??), where $f_1$ and $f_2$ are given by the variables `f1` and `f2`, respectively. It is also known that each of the signals $m_1(t)$, $m_2(t)$, and $m_3(t)$ correspond to a single letter of the alphabet which has been encoded using International Morse Code, as shown in the following table:

| A | · - | H | ···· | O | - - - | V | ··· - |
|---|---|---|---|---|---|---|---|
| B | - ··· | I | ·· | P | · - - · | W | · - - |
| C | - · - · | J | · - - - | Q | - - · - | X | - ·· - |
| D | - ·· | K | - · - | R | · - · | Y | - · - - |
| E | · | L | · - ·· | S | ··· | Z | - - ·· |
| F | ·· - · | M | - - | T | - | | |
| G | - - · | N | - · | U | ·· - | | |

### Basic Problems

(a). Using the signals dot and dash, construct the signal that corresponds to the letter 'Z' in Morse code, and plot it against t. As an example, the letter C is constructed by typing c = [dash dot dash dot]. Store your signal $z(t)$ in the vector z.

(b). Plot the frequency response of the filter using freqs(bf,af).

(c). The signals dot and dash are each composed of low frequency components such that their Fourier transforms lie roughly within the passband of the lowpass filter. Demonstrate this by filtering each of the two signals, using

```
>> ydash=lsim(bf,af,dash,t(1:length(dash)));
>> ydot=lsim(bf,af,dot,t(1:length(dot)));
```

Plot the outputs ydash and ydot along with the original signals dash and dot.

(d). When the signal dash is modulated by $\cos(2\pi f_1 t)$, most of the energy in the Fourier transform will move outside the passband of the filter. Create the signal $y(t)$ by executing y=dash.*cos(2*pi*f1*t(1:length(dash))). Plot the signal $y(t)$. Also plot the output yo=lsim(bf,af,y,t). Do you get a result that you would have expected?

### Intermediate Problems

(e). Determine analytically the Fourier transform of each of the signals

$$m(t)\cos(2\pi f_1 t)\cos(2\pi f_1 t),$$

$$m(t)\cos(2\pi f_1 t)\sin(2\pi f_1 t),$$

and

$$m(t)\cos(2\pi f_1 t)\cos(2\pi f_2 t)$$

in terms of $M(j\omega)$, the Fourier transform of $m(t)$.

(f). Using your results from Part ?? and by examining the frequency response of the filter as plotted in Part ??, devise a plan for extracting the signal $m_1(t)$ from $x(t)$. Plot the signal $m_1(t)$ and determine which letter is represented in Morse code by the signal.

(g). Repeat Part ?? for the signals $m_2(t)$ and $m_3(t)$. Agent 008, where does the future of technology lie?

## ■ 4.7 Symbolic Computation of the Continuous-Time Fourier Transform Ⓢ

This exercise uses the Symbolic Math Toolbox to evaluate the continuous-time Fourier transform (CTFT) given by Eq. (??) for several different signals.

## Basic Problems

(a). Define symbolic expressions **x1** and **x2** to represent the following continuous-time signals:

$$x_1(t) = (1/2)e^{-2t}u(t)$$
$$x_2(t) = e^{-4t}u(t).$$

You will need to use the function **Heaviside** to represent the unit step function $u(t)$, as in

```
>> x1=sym('(1/2)*exp(-2*t)*Heaviside(t)');
>> x2=sym('exp(-4*t)*Heaviside(t)');
```

(b). For $x_1(t)$ and $x_2(t)$ defined in Part **??**, analytically compute the value of their CTFTs evaluated at $\omega = 0$, i.e., $X(j\omega)|_{\omega=0}$. You should not compute $X(j\omega)$ to do this. How is the CTFT at $\omega = 0$ related to the time-domain signal?

(c). Which of the signals defined in Part **??** decays faster in the time domain? Based on this, which do you expect to decay more rapidly in frequency?

(d). Use the function **fourier** to compute the CTFTs of $x_1(t)$ and $x_2(t)$. Define **X1** and **X2** to be the symbolic expressions returned by **fourier**. Use **ezplot** to generate plots of the magnitudes of $X_1(j\omega)$ and $X_2(j\omega)$. Do these plots confirm your answers to Parts **??** and **??**?

## Intermediate Problems

(e). Define the symbolic expression **y1** to represent the continuous-time signal

$$y_1(t) = \begin{cases} 1, & -2 \le t \le 2, \\ 0, & \text{otherwise}, \end{cases}$$

as the difference of two **Heaviside** functions.

(f). Analytically determine $Y_1(j\omega)$, the CTFT of $y_1(t)$.

(g). Define the symbolic expression **y2** to represent the signal $y_2(t) = y_1(t-2)$. You can do this by using the difference of two **Heaviside** functions as you did for **y1**, or by an appropriate use of **subs** on **y1**.

(h). Use **fourier** to find the CTFTs of **y1** and **y2** and store the results in **Y1** and **Y2**. If **Y1** is not the expression you expected to get, then try using **simple** on this expression to put it in a more familiar form.

(i). Generate plots of the magnitudes of $Y_1(j\omega)$ and $Y_2(j\omega)$ with **ezplot**. How do these plots compare? Could you have predicted this from the relationship between these signals in the time domain?

(j). In the next few parts, you will find the CTFT for the signal

$$v(t) = e^{-2|t|}.$$

Write $v(t)$ as the sum of two signals $v_1(t)$ and $v_2(t)$. Choose $v_1(t)$ to be a causal signal, and $v_2(t)$ to be an anticausal signal, i.e., $v_2(t) = 0$ for $t > 0$. Analytically compute $V(j\omega)$, the CTFT of $v(t)$.

(k). Define symbolic expressions `v1` and `v2` for the signals you defined in Part ??, and then combine them using `v=v1+v2` to get an expression for the original signal $v(t)$.

(l). Use `fourier` to find a symbolic expression `V` for the CTFT of `v`. Is this expression equivalent to the one you found analytically in Part ???

## Advanced Problems

(m). Define `f` to be a symbolic expression for the signal $f(t) = e^{-at}u(t)$. Use `fourier` to define `F` to be the symbolic expression for the CTFT of `f`. Note that `F` contains an unevaluated expression. Does the unevaluated expression converge for all values of $a$?

(n). Use `subs` to set the value of $a$ in `F` to be 5, and then apply `simple` to the result of the substitution. Does this give the result you expect?

Chapter 5

# The Discrete-Time Fourier Transform

In Chapter ??, you learned why the continuous-time Fourier series (CTFS) and the discrete-time Fourier series (DTFS) are useful tools for analyzing and manipulating periodic signals. These frequency-domain tools were extended to aperiodic continuous-time signals with the continuous-time Fourier transform (CTFT), which was the subject of Chapter ??. The exercises in this chapter cover the discrete-time Fourier transform (DTFT) and are designed to complete your introduction to Fourier analysis. While the DTFS can only represent periodic signals, the DTFT is a unified framework for representing both periodic and aperiodic discrete-time signals in terms of complex exponential signals $e^{j\omega n}$, just as the CTFT is a unified framework for representing continuous-time signals in terms of complex exponential signals $e^{j\omega t}$. The DTFT of a discrete-time signal $x[n]$ is given by

$$X(e^{j\omega}) = \sum_{n=-\infty}^{\infty} x[n]e^{-j\omega n}, \tag{5.1}$$

which is called the DTFT analysis equation. The synthesis equation is given by

$$x[n] = \frac{1}{2\pi}\int_{2\pi} X(e^{j\omega})e^{j\omega n}d\omega, \tag{5.2}$$

where the interval of integration can be any interval of length $2\pi$. In this chapter, you will to learn to implement numerically the DTFT analysis and synthesis equations using the functions `fft` and `ifft`, respectively. The two major limitations for any numerical implementation of the DTFT are that $x[n]$ must have finite length and $X(e^{j\omega})$ can only be computed at a finite number of samples of the continuous frequency variable $\omega$. These limitations are addressed in Exercise ??. Exercise ?? shows why the DTFT is a useful tool for analyzing signals arising in practical applications, such as the tones transmitted by a touch-tone phone. The DTFT is also used to analyze discrete-time LTI systems. Systems for which the frequency response magnitude is equal to one at all frequencies, known as all-pass systems, are examined in Exercise ??. Exercises ?? and ?? show how the DTFT can be used both to design and identify LTI systems. The final exercise of this chapter, Exercise ??, demonstrates how the function `residue` can be used to compute the partial fraction expansion of frequency responses which are rational polynomials in $e^{-j\omega}$. These expansions can then be used to derive analytic expressions for the impulse responses.

## 5.1 Computing Samples of the DTFT

This exercise will examine the computation of the discrete-time Fourier transform (DTFT) in MATLAB. A fundamental difference between the DTFT and the CTFT is that the DTFT is periodic in frequency. Mathematically, this can be shown by examining the DTFT equation,

$$\begin{aligned} X\left(e^{j(\omega+2\pi)}\right) &= \sum_{n=-\infty}^{\infty} x[n]e^{-j(\omega+2\pi)n}, \\ &= \sum_{n=-\infty}^{\infty} x[n]e^{-j\omega n}(1)^n, \\ &= X(e^{j\omega}). \end{aligned}$$

The DTFT is periodic with period $2\pi$ because constraining $n$ to be an integer means that the complex exponentials $e^{j\omega n}$ are only distinct over a range of $\omega$ of length $2\pi$. The principal period, $-\pi \leq \omega < \pi$ is usually chosen, although any interval of length $2\pi$ will do.
There are two issues which must be addressed before computing the DTFT of a signal using MATLAB. First, if $x[n]$ is an infinite-length signal, then $x[n]$ must be truncated to a finite-length signal, since only finite-length signals can be represented by vectors in MATLAB. Another issue of practical importance is that $X(e^{j\omega})$ is defined over the continuous variable $\omega$. However, $X(e^{j\omega})$ can only be computed at a discrete set of frequency samples, $\omega_k$. If enough frequency samples are chosen, then a plot of these frequency samples will be a good representation of the actual DTFT. For computational efficiency, the best set of frequency samples is the set of equally spaced points in the interval $0 \leq \omega < 2\pi$ given by $\omega_k = 2\pi k/N$ for $k = 0, \ldots, N-1$. For a signal $x[n]$ which is nonzero only for $0 \leq n \leq M-1$, these frequency samples correspond to

$$X(e^{j\omega_k}) = \sum_{n=0}^{M-1} x[n]e^{-j2\pi kn/N}, \quad k = 0, \ldots, N-1. \tag{5.3}$$

The function `fft` implements Eq. (??) in a computationally efficient manner. If `x` is a vector containing $x[n]$ for $0 \leq n \leq M-1$ and $N \geq M$, then `X=fft(x,N)` computes $N$ evenly spaced samples of the DTFT of `x` and stores these samples in the vector `X`. If $N < M$, then the MATLAB function `fft` truncates `x` to its first $N$ samples before computing the DTFT, thus yielding incorrect values for the samples of the DTFT. You will see how to compute fewer than $M$ frequency samples in one of the following problems.
The values of `X(k)` for $k \geq N/2$ are also samples of $X(e^{j\omega})$ on the interval $-\pi \leq \omega < 0$, due to the periodicity of the DTFT. If you want to reorder the DTFT samples returned by `fft` to correspond to the interval $-\pi \leq \omega_k < \pi$, the MATLAB function `fftshift` was written specifically to swap the second half of the vector `X` with the first half.

### Basic Problems

(a). Calculate analytically the DTFT of the rectangular pulse defined by $x[n] = u[n] - u[n-11]$. Also create the vector `x` containing the nonzero samples of $x[n]$.

(b). Create a vector of $N = 100$ frequencies containing the frequency samples w=2*pi*k/N for k=[0:N-1]. Plot $|X(e^{j\omega})|$ over this range, using the formula you calculated in Part ??. Also plot the phase $\angle X(e^{j\omega})$. You may find the following function helpful:

```
function X=dtftsinc(M,w)
% X=dtftsinc(M,w)
% calculates the function, X = sin(Mw/2)/sin(w/2)
% using selective indexing to avoid division by 0
den=sin(w/2);
num=sin(M*w/2);
X=zeros(size(w));
X(den~=0)=num(den~=0)./den(den~=0);
X(den==0)=M;
```

This function is provided in the Computer Explorations Toolbox, or you can create a new M-file for it.

(c). Rearrange the frequency samples to correspond to the principle period of the DTFT, $-\pi \leq \omega < \pi$. To do this, use w=w-pi. Use the function fft to calculate N=100 samples of the DTFT of $x[n]$ and store your result in the vector X. Plot the magnitude and phase of X versus w. Don't forget to use fftshift to rearrange the DTFT samples in X to match the frequencies in w. How does this plot compare to your results from Part ???

(d). Use the time-shift property of the DTFT to calculate analytically the DTFT of the signal $x[n+5]$, i.e., the centered pulse. You should get an answer that is purely real. Again, using the time-shift property of the DTFT, determine the value of the parameter a so that Xr=exp(j*w*a).*X corresponds to the DTFT of $x[n+5]$. Plot Xr versus w. You may have to use the function real to remove any small imaginary parts of your result. Also use the function dtftsinc along with your analytic expression for the DTFT of $x[n+5]$ to verify your answer.

(e). Use fft to compute the DTFT of the signal $z[n] = (5-|n|)(u[n+5]-u[n-5])$. Plot the DTFT of $z[n]$ versus w. Hint: The DTFT of $z[n]$ should be a purely real function of $\omega$, so real may be used to eliminate any small imaginary parts which result from numerical round-off errors.

## Intermediate Problems

In the following problems, you will learn how to calculate $N$ samples of $X(e^{j\omega})$, where $N$ is less than the length of $x[n]$. The function fft(x,N) truncates the vector x to N samples whenever length(x)>N. To see that this is not the correct way to get $N$ samples of the DTFT of x, consider X=fft(x,1). Here, only one frequency sample of $X(e^{j\omega})$ is desired, $X(e^{j0})$, the DC component of x. Examination of the definition of the DTFT, Eq. (??), reveals that $X(e^{j0}) = \sum_{n=0}^{M-1} x[n]$. However, fft(x,1) simply returns x(1), the first sample of x.

(f). Given that the function `fft(x,N)` truncates the vector `x` to `N` samples, whenever `length(x)>N`, determine what MATLAB will return for the following call, `Y=fft(x,2)`, where `x` is an arbitrary vector of length 100. Your answer should only involve the first two samples of the vector `x`. To determine this expression, you might try `x=[1:100]` and `x=[1 zeros(1,99)]`.

(g). If only two samples of the DTFT are desired, $\omega_k = 2\pi k/2$, for $k = 0, 1$, how can they be computed in MATLAB without using `fft`? Calculate these two frequency samples, $\omega_k = 0, \pi$, for the signal `x=[1:10 10:-1:1]`.

(h). Assume for this part that $N = M$. Consider the DTFT equation, evaluated at $\omega_k = 2\pi(2k)/N$, i.e., for the even values $2k$. These $N/2$ samples of the DTFT of $x[n]$ are given by

$$X(e^{j\omega_k}) = \sum_{n=0}^{M-1} x[n]e^{-j2\pi kn/(N/2)}, \quad k = 0, \ldots, N/2 - 1. \tag{5.4}$$

Now define a signal $g[n]$ in terms of $x[n]$ such that the DTFT of $x[n]$ can be rewritten as a sum over the range $n = 0, \ldots, M/2 - 1$ as

$$X(e^{j\omega_k}) = \sum_{n=0}^{M/2-1} g[n]e^{-j2\pi kn/(N/2)}, \quad k = 0, \ldots, N/2 - 1. \tag{5.5}$$

Use this result to calculate 100 evenly spaced samples of the DTFT of the signal $x[n] = (0.975)^n \cos(0.3n)(u[n] - u[n - 200])$. Plot the magnitude and phase of the DTFT versus `w`.

## Advanced Problems

Rather than calculating $M/2$ samples, you may wish to calculate $N$ of samples of the DTFT of $x[n]$ for any $N < M$, where `M=length(x)`. The algorithm for obtaining these frequency samples is based upon the following observation: if a sequence of length $N$, $\tilde{x}[n]$, can be found having a DTFT which is equal to $X(e^{j\omega})$ at the desired frequency samples, then `fft` can be used.

(i). Consider a sequence $x[n]$ with DTFT $X(e^{j\omega})$. Show that the frequency samples

$$\tilde{X}[k] = X(e^{j\omega})\big|_{\omega=(2\pi/N)k} = X\left(e^{j(2\pi/N)k}\right) \tag{5.6}$$

are the same as the frequency samples of $\tilde{X}(e^{j\omega})$ which is the DTFT of $\tilde{x}[n]$. The signal $\tilde{x}[n]$ is the length $N$ sequence

$$\tilde{x}[n] = \begin{cases} \sum_{r=-\infty}^{\infty} x[n + rN], & 0 \le n \le N - 1, \\ 0, & \text{otherwise}. \end{cases} \tag{5.7}$$

The signal $\tilde{x}[n]$ is called a "time-aliased" version of $x[n]$.

(j). Create a MATLAB function that takes as input a sequence `x` of arbitrary length and calculates $N$ equally spaced frequency samples of its DTFT. The first line of your function should be

```
function X=dtft(x,N)
```

Be sure that your function works properly for both the case when $N \geq M$, and $N < M$. You should use exactly one call to the MATLAB function `fft` for each input `x`.

## 5.2 Telephone Touch-Tone

This exercise will teach you how the touch-tone system on the telephone uses signals of different frequencies to indicate which key has been pushed. The DTFT of a sampled telephone signal can be used to identify these frequencies. The sound you hear when you push a key is the sum of two sinusoids. The higher frequency sinusoid indicates the column of the key on the touch-pad and the lower frequency sinusoid indicates the row of the key on the touch-pad. Figure ?? shows the layout of a telephone keypad and the two DTFT frequencies corresponding to each digit, assuming the continuous-time waveform is sampled at 8192 kHz. The figure also contains a table listing each digit and the DTFT frequencies for that digit. For example, the digit 5 is represented by the signal

$$d_5[n] = \sin(0.5906\, n) + \sin(1.0247\, n). \tag{5.8}$$

| | $\omega_{column}$ | | |
|---|---|---|---|
| $\omega_{row}$ | 0.9273 | 1.0247 | 1.1328 |
| 0.5346 | 1 | 2 | 3 |
| 0.5906 | 4 | 5 | 6 |
| 0.6535 | 7 | 8 | 9 |
| 0.7217 | | 0 | |

| Digit | $\omega_{row}$ | $\omega_{column}$ |
|---|---|---|
| 0 | 0.7217 | 1.0247 |
| 1 | 0.5346 | 0.9273 |
| 2 | 0.5346 | 1.0247 |
| 3 | 0.5346 | 1.1328 |
| 4 | 0.5906 | 0.9273 |
| 5 | 0.5906 | 1.0247 |
| 6 | 0.5906 | 1.1328 |
| 7 | 0.6535 | 0.9273 |
| 8 | 0.6535 | 1.0247 |
| 9 | 0.6535 | 1.1328 |

**Figure 5.1.** DTFT frequencies for touch-tone signals sampled at 8192 Hz.

### Basic Problems

In these problems, you will create the appropriate touch-tone for each digit, and examine the DTFT to make sure the signals have the correct frequencies. You will also define a vector containing the touch-tones for your phone number.

(a). Create row vectors `d0` through `d9` to represent all 10 digits for the interval $0 \leq n \leq 999$. Listen to each signal using `sound`. For example, `sound(d2,8192)` should sound like the tone you hear when you push a '2' on the telephone.

(b). The function `fft` can be used to compute $N$ samples of the DTFT of a finite-length signal at frequencies $\omega_k = 2\pi k/N$. For example, `X = fft(x,2048)` computes 2048 evenly spaced samples of $X(e^{j\omega})$ at $\omega_k = 2\pi k/2048$ for $0 \leq k \leq 2047$. Use `fft` to compute samples of $D_2(e^{j\omega})$ and $D_9(e^{j\omega})$ at $\omega_k = 2\pi k/2048$. Define `omega` to be a vector containing $\omega_k$ for $0 \leq k \leq 2047$. Plot the magnitudes of the DTFTs for these signals, and confirm that the peaks fall at the frequencies specified in Figure ??. You will find it easier to see which frequencies are present if you use `axis` to restrict the $\omega$-axis to the region $0.5 \leq \omega \leq 1.25$ while keeping the full vertical range. Generate appropriately labeled plots of the DTFT magnitudes for these two digits.

(c). Define `space` to be a row vector with 100 samples of silence using `zeros`. Define `phone` to be your phone number by appending the appropriate signals and `space`. For instance, if your phone number was 555-7319, you would type

```
>> phone = [d5 space d5 space d5 space d7 space d3 space d1 space d9];
```

Note that in order to append the signals this way, all of the digits you defined in Part ?? and `space` must be row vectors. Play your phone number using `sound` and verify that it sounds the same as when you dial it on a touch-tone phone.

## Intermediate Problems

For these problems, you will learn to decode phone numbers from their touch-tones. The phone numbers to be decoded are in a file called `touch.mat`, which is in the Computer Explorations Toolbox. The Computer Explorations Toolbox can be obtained from The MathWorks at the address listed in the Preface. The data can be loaded into MATLAB by typing `load touch` as long as the file is in your MATLABPATH. If the file loaded correctly, you should be able to list the names of variables by typing

```
>> who

Your variables are:

hardx1    hardx2    x1    x2
```

The vectors `x1` and `x2` contain sampled versions of the sequence of touch-tones representing two different phone numbers. As in Part ??, the signals consist of 7 digits of 1000 samples each, separated by 100 samples of silence. The vectors `hardx1` and `hardx2` are less precisely dialed versions of the same numbers that are used in Part ??.

(d). Using `fft`, compute 2048 evenly spaced samples of the DTFT for each digit of `x1`. In order to apply `fft` to each digit separately, you will need to segment the signal into seven digits using the information you have about the relative lengths of the digits and spaces, or by plotting the signal and identifying when each digit starts and stops. Apply `fftshift` to the output of `fft`, which will rearrange the samples of the DTFT so they correspond to ascending frequencies in the range $-\pi \leq \omega_k < \pi$. Define `X11` through `X17` to contain these samples of the DTFT you obtained by applying `fftshift` to the output of `fft` for each digit. To determine the digits of the telephone number represented by `x1`, plot the magnitude of the DTFTs and compare the peak frequencies of the signals with those shown in Figure ?? As a check for your answer, the sum of the digits should be 41.

(e). Repeat Part ?? for the signal `x2`, and decode the digits of this phone number as well. These digits may not sum to 41.

## Advanced Problems

In these problems, you will write a function to decode phone numbers automatically from the touch-tones. To help design your decoder, you will look at the energy of the tones at each of the possible frequencies from Figure ??.

(f). Using `fft` to compute 2048 samples of $X(e^{j\omega})$, figure out which value of $\omega_k$, and the corresponding index `k`, is closest to each of the touch-tone frequencies. Remember that MATLAB vectors begin with index `k=1`, so the DTFT sample at $\omega = 0$ is stored in `X(1)`.

(g). The value of $|X(e^{j\omega_k})|^2$ gives the energy in a signal at frequency $\omega_k$. Apply `fft` to `d8`, which was defined in Part ??, and use the output of `fft` to compute $|D_8(e^{j\omega_k})|^2$ for each of the $\omega_k$ you determined in Part ??. Is the energy highest for appropriate values of $\omega_k$ corresponding to an '8'?

(h). Write a function `ttdecode` which accepts as its input a touch-tone signal in the format used in Part ??, (1000 samples of touch-tone for each digit, separated by 100 samples of silence), and returns as its output a seven element vector containing the phone number. For example, if the vector `phone` contained the touch-tones for the number 555-7319, you should get the following output:

```
>> testout = ttdecode(phone);
>> testout
testout =
5 5 5 7 3 1 9
```

The first line of the M-file you write to implement the function should read

```
function digits = ttdecode(x)
```

Your function should use `fft` to compute 2048 samples of the DTFT of each digit in `x`, and then check the energy at the samples you specified in Part ?? corresponding to the $\omega_k$ for the touch-tones. Pick which of the row frequencies and column frequencies has the most energy, then use those frequencies to determine what the digit is. Test your function by using x1 and x2 as inputs and verify that it returns the same phone number as you obtained in Parts ?? and ??. Also, test your function on the signal you created in Part ?? to represent your own phone number.

(i). Most people do not dial their telephone with the ruthless precision assumed in this exercise, with each digit exactly the same length and the space between digits always the same length. Modify your function to work with touch-tone signals where the digits and silences can have varying lengths. To simplify things slightly, assume that all the tones and silences are at least 100 samples long. Verify the new version of your function works with the signals hardx1 and hardx2. These signals contain the same tones as x1 and x2, but they are not regularly spaced.

## 5.3 Discrete-Time All-Pass Systems

This exercise explores the effect of all-pass systems on discrete-time signals. All-pass systems are defined to have a frequency response with magnitude equal to one, i.e., $|H(e^{j\omega})| = 1$. The magnitude of the DTFT of the output signal of an all-pass system is identical to the magnitude of the DTFT of the input signal. However, as will be demonstrated in this exercise, the phase of the frequency response can lead to significant distortion in the output of the system.

You will work with two different discrete-time all-pass systems in this exercise. Both of these systems are causal LTI systems. Their inputs and outputs satisfy the following linear, constant-coefficient difference equations:

$$\text{System 1:} \qquad y[n] = x[n-3]\,,$$

$$\text{System 2:} \qquad y[n] - (3/4)y[n-1] = -(3/4)x[n] + x[n-1]\,.$$

### Basic Problems

(a). Define coefficient vectors a1 and b1 for System 1 and use freqz to generate appropriately labeled plots of the magnitude and phase of the frequency response $H_1(e^{j\omega})$ of this system. If you are unfamiliar with freqz, you may wish to review Tutorial ??. Do your plots confirm that the system is an all-pass system?

(b). Define coefficient vectors a2 and b2 for System 2 and generate plots of the magnitude and phase of the frequency response of this system. Again, your plots should show that $|H_2(e^{j\omega})| = 1$ for all $\omega$. Is the phase of $H_2(e^{j\omega})$ the same as the phase of $H_1(e^{j\omega})$? When the inputs to both systems are the same, would you expect the outputs to be the same?

### Intermediate Problems

(c). Define x to be the signal $x[n] = (3/4)^n\, u[n]$ for $0 \le n \le 50$. Create an appropriately labeled plot of x using stem.

(d). Let $y_1[n]$ and $y_2[n]$ be the outputs of Systems 1 and 2 respectively when the input is $x[n]$. Use the function filter described in Tutorial ?? to compute the vectors y1 and y2 containing $y_1[n]$ and $y_2[n]$ for $0 \leq n \leq 50$. Use `subplot` to make a figure showing both outputs. Are the outputs the same?

(e). Use `fft` as described in Exercise ?? to compute 1024 samples of $X(e^{j\omega})$, $Y_1(e^{j\omega})$, and $Y_2(e^{j\omega})$, the DTFTs of the input and both outputs. Place these samples in `X`, `Y1`, and `Y2`. Plot the magnitudes of all three vectors. How do they compare? How could you have anticipated this result?

(f). Which of the two outputs looks most like the input $x[n]$? Does preserving the magnitude of the DTFT of a signal guarantee the signal is not distorted in the time domain?

## Advanced Problems

(g). Consider using the output of System 2 computed in Part ?? as the input of another system equivalent to System 2. Define $y_{22}[n]$ to be the corresponding output, i.e., $y_{22}[n]$ is the output of a system corresponding to a cascade of two versions of System 2 when the input is $x[n] = (3/4)^n u[n]$. Will $|Y_{22}(e^{j\omega})| = |X(e^{j\omega})|$? In general, if you cascade M versions of System 2, will the output of this cascade have the same DTFT magnitude as the input?

(h). Consider the systems obtained by cascading 2, 3 and 4 copies of System 2. Define vectors `y22`, `y23` and `y24` to represent the outputs of these systems computed using `filter`. Plot `x`, `y2`, and each of the three outputs you just calculated on the same set of axes. How would you describe the effect of each additional all-pass?

(i). As you have seen, an all-pass system may significantly distort the appearance of a signal in the time domain without changing the signal energy at each frequency. It is important to understand the effect of a system on the energy in a signal as well as the appearance in the time domain. Compute the energy in `x` as $\sum_{n=0}^{50} x^2[n]$. Repeat this calculation for `y2`, `y22`, `y23` and `y4`. How do the energies of these signals compare? Could you have predicted this?

## ■ 5.4 Frequency Sampling: DTFT-Based Filter Design

The frequency response of any stable LTI filter is given by the DTFT of its impulse response. Because MATLAB can be used to compute samples of the DTFT of finite-length signals, as shown in Exercise ??, MATLAB can also be used to compute samples of the frequency response of finite-length impulse response (FIR) filters. In this exercise, you will use the DTFT to design an FIR filter from samples of the desired frequency response. Since this method of filter design only specifies the frequency response at a set of frequency samples, it is often called "frequency sampling".

The frequency response of an FIR filter whose impulse response $h[n]$ is nonzero only on the interval $0 \leq n \leq N-1$ is given by

$$H(e^{j\omega}) = \sum_{n=0}^{N-1} h[n]e^{-j\omega n}. \tag{5.9}$$

Exercise ?? demonstrates how to use the function `fft` to compute $N$ equally spaced samples of the DTFT. In this set of problems, you will specify the values of the DTFT at these $N$ frequency values, and then will use the operator \ to solve the set of linear equations that result. The FIR filter will be an approximation to the ideal lowpass filter with cutoff frequency $\omega_c = \pi/2$:

$$H_{id}(e^{j\omega}) = \begin{cases} 1, & |\omega| < \pi/2, \\ 0, & \pi/2 \leq |\omega| \leq \pi. \end{cases} \tag{5.10}$$

## Basic Problems

(a). Draw the desired frequency response magnitude over the range $0 \leq \omega \leq 2\pi$. Remember that the DTFT is periodic with period $2\pi$.

(b). Construct the vector corresponding to the $N = 9$ equally spaced frequencies from 0 to $2\pi$, `w=2*pi*k/N`, for `N=9` and `k=[0:N-1]`. Store in a vector `Hm` the desired frequency response magnitudes $|H_{id}(e^{j\omega})|$ that correspond to the frequencies in the vector `w`. Plot the desired frequency response magnitude versus `w`, using `plot(w,Hm)`. Does your plot look like a lowpass filter?

(c). Now that the frequency response magnitude has been specified, you need to specify the phase. The frequency response $H(e^{j\omega})$ could be made purely real if $h[n]$ were symmetric about $n = 0$. In this case, $H(e^{j\omega})$ would have zero phase. However, as assumed by Eq. (??), the impulse response you are to compute must be causal, and thus cannot be symmetric about $n = 0$. What is the phase of the causal filter $h[n]$, which is related to the zero-phase filter by a delay of $(N-1)/2$ samples? Store in the vector `Hp` the values of this phase at the frequency samples contained in `w`. Hint: The phase should be of the form $a\omega$ for frequency samples $|\omega| \leq \pi$. Use the periodicity of $H(e^{j\omega})$ to determine the phase samples in `Hp` for $\omega > \pi$.

## Intermediate Problems

(d). The two vectors you just created, `Hm` and `Hp`, contain nine pieces of information about the nine-point impulse response $h[n]$. Each of the samples of the DTFT of $h[n]$ can be stated in the form of a linear equation involving $h[n]$. By simultaneously solving these nine linear equations for the nine unknowns, $h[n], n = 0, \ldots, 8$, you can recover the impulse response of the lowpass filter. When written in matrix form, these equations are

$$\underbrace{\begin{bmatrix} 1 & e^{-j\omega_0} & e^{-j2\omega_0} & \ldots & e^{-j8\omega_0} \\ 1 & e^{-j\omega_1} & e^{-j2\omega_1} & \ldots & e^{-j8\omega_1} \\ \vdots & \vdots & \vdots & \vdots & \vdots \\ 1 & e^{-j\omega_8} & e^{-j2\omega_8} & \ldots & e^{-j8\omega_8} \end{bmatrix}}_{\texttt{A}} \underbrace{\begin{bmatrix} h[0] \\ h[1] \\ \vdots \\ h[8] \end{bmatrix}}_{\texttt{h}} = \underbrace{\begin{bmatrix} H(e^{j\omega_0}) \\ H(e^{j\omega_1}) \\ \vdots \\ H(e^{j\omega_8}) \end{bmatrix}}_{\texttt{H}}, \tag{5.11}$$

where $\omega_k = 2\pi k/9$. The matrix on the left hand side of Eq. (??) is known as a DFT matrix and can be created with the call `A=exp(-j*2*pi*[0:8]'*[0:8]/9)`. Create the desired frequency response vector by storing `Hm.*exp(j*Hp)` in the column vector `H`, i.e., `H=Hm(:).*exp(j*Hp(:))`. You can now use the operator \ to solve this set of simultaneous linear equations by typing `h=A\H`. Plot the resulting impulse response using `stem`. The impulse response should be real, although you may need to use `real` to remove an imaginary part on the order of `10e-16` due to roundoff errors.

(e). Verify that the impulse response has the desired DTFT magnitude at the specified frequency samples by plotting the magnitude of the DTFT of $h[n]$ and the magnitude of the desired DTFT samples on the same plot. To do this, first create a vector `w2` to contain 1000 evenly spaced samples of the frequency interval $0 \leq \omega < 2\pi$. Then calculate the DTFT of `h` at these samples using `fft` and store the result in the vector `H2`. Now plot `abs(H2)` versus `w2` and `H` versus `w` on the same plot using, `plot(w2,abs(H2),'-',w,Hm,'o')`. Be sure to plot at least 1000 samples of the DTFT magnitude to see what happens between the samples you specified.

## 5.5 System Identification

Linear time-invariant (LTI) systems have been particularly successful in modeling numerous physical processes. While an LTI system often can be derived directly from physical laws, in many cases the model must be deduced from external observations. Such observations might include measuring the response to a known input $x[n]$. If the input were an impulse, then the response $y[n]$ would be the impulse response $h[n]$, which completely determines the LTI system. For most applications, however, one does not have complete control over the input signal. In this case, the convolution property of the DTFT can be used to solve for the impulse response. That is, if the DTFTs of the system input and output can be computed, then

$$H(e^{j\omega}) = \frac{Y(e^{j\omega})}{X(e^{j\omega})} \tag{5.12}$$

provides the frequency response, which is the DTFT of the impulse response. In MATLAB, the DTFT of a finite-length signal can be computed at evenly spaced frequencies on the interval $0 \leq \omega < 2\pi$ using `fft`. Namely, if `x` is a vector containing $x[n]$ on the interval $0 \leq n \leq N-1$, then `X=fft(x)` is a vector containing $X(e^{j\omega_k})$ at $\omega_k = 2\pi k/N$ for $0 \leq k \leq N-1$. Since samples of $Y(e^{j\omega})$ can be computed similarly, $H(e^{j\omega})$ can be derived for these frequency samples. The function `ifft` can then be invoked to compute $h[n]$. Namely, if $h[n]$ is a finite-length signal and is nonzero outside the interval $0 \leq n \leq N-1$, then `ifft(H)` returns the values of $h[n]$ on this interval when `H` is a vector containing the aforementioned frequency samples of $H(e^{j\omega})$. For a more in-depth discussion of the use of `fft` and `ifft`, see Exercise **??**.

### Basic Problems

Suppose that the response of a particular LTI system to the input $x[n] = (-3/4)^n\, u[n]$ were

$$y[n] = \frac{2}{5}\left(\frac{1}{2}\right)^n u[n] + \frac{3}{5}\left(-\frac{3}{4}\right)^n u[n]\,.$$

You are to determine the impulse response of this system using `fft` and `ifft`.

(a). Store in `x` and `y` the values of $x[n]$ and $y[n]$ on the interval $0 \leq n \leq N-1$, where $N = 64$. Plot these signals to convince yourself that they are essentially equal to zero outside this interval, and thus can be safely truncated.

(b). Use `fft` to compute $X(e^{j\omega})$ and $Y(e^{j\omega})$ at $N = 64$ evenly spaced frequencies on the interval $0 \leq \omega < 2\pi$. Plot $|X(e^{j\omega})|$ and $|Y(e^{j\omega})|$ versus $\omega$ on the interval $0 \leq \omega < 2\pi$.

(c). Compute $H(e^{j\omega})$ at the same 64 frequency samples. Use `ifft` to compute $h[n]$ on the interval $0 \leq n \leq N-1$.

(d). Because both the system input and output have DTFTs which are ratios of polynomials of $e^{-j\omega}$, the impulse response can also be determined analytically. Determine analytical expressions for $Y(e^{j\omega})$ and $X(e^{j\omega})$, and then determine analytically the inverse DTFT of $Y(e^{j\omega})/X(e^{j\omega})$. Use this impulse response to to verify the impulse response computed in Part ??.

Note that there are essentially two methods for computing the impulse response of a discrete-time LTI system in MATLAB. First, if the frequency response of the system can be expressed as a ratio of polynomials in $e^{-j\omega}$, then `residue` can be used to compute the partial fraction expansion of the frequency response. The partial fraction expansion allows you to then derive the impulse response, which can have infinite length. If the impulse response has finite length, then the MATLAB function `ifft` can be used to compute $h[n]$ from samples of the frequency response. The advantage of the second method is that the frequency response need not be expressible as a ratio of polynomials in $e^{-j\omega}$.

## Intermediate Problems

Now consider finding the impulse response of an LTI system whose output is

$$y[n] = -\frac{2}{5}\left(\frac{1}{2}\right)^n u[n] + \frac{7}{5}\left(-\frac{3}{4}\right)^n u[n] \tag{5.13}$$

when the input is $x[n] = (1/2)^n u[n] - (1/2)^{n-1} u[n-1]$.

(e). Store in `x` and `y` the values of $x[n]$ and $y[n]$ on the interval $0 \leq n \leq N-1$, where $N = 64$.

(f). Use `fft` to compute $X(e^{j\omega})$ and $Y(e^{j\omega})$ at $\omega_k = 2\pi k/N$ for $0 \leq k \leq N-1$. Plot $|X(e^{j\omega})|$ and $|Y(e^{j\omega})|$ versus $\omega$. Are there any frequencies for which $|X(e^{j\omega})|$ is essentially equal to zero?

(g). Using the results of Part ?? and Eq. (??), compute $H(e^{j\omega})$ at the frequencies $\omega_k = 2\pi k/N$, $0 \leq k \leq N-1$. Store these frequency response values in the vector `H`. Next use `ifft` to compute $h[n]$ on the interval $0 \leq n \leq N-1$. Store this impulse response in the vector `h1`.

(h). Verify that the impulse response stored in `h1` is correct by computing the convolution `y1=conv(h1,x)`. How does `y1` compare with the known system response on the interval $0 \leq n \leq N-1$? Can you explain this discrepancy? Hint: Is there any way to determine the value of $H(e^{j\omega_k})$ when $X(e^{j\omega_k}) = 0$? Remember that the values of the DTFT computed by `fft` are equal to $X(e^{j\omega})$ only if $x[n]$ is truly zero outside the interval $0 \leq n \leq N-1$; thus $X(e^{j\omega_k})$ might be zero for $\omega = \omega_k$, but the value computed by `fft` is nonzero (and very small) due to approximation errors.

To overcome the problems introduced by zero values of $X(e^{j\omega})$, you can make some assumptions about the behavior of the frequency response near frequencies for which $X(e^{j\omega}) = 0$. One common assumption is that $H(e^{j\omega})$ is a smoothly varying function.

(i). Find the frequencies at which $X(e^{j\omega})$ is essentially zero. "Essentially zero" means that the value of $|X(e^{j\omega})|$ computed by `fft` is extremely small and likely due to approximation errors. Create a new vector `H2` equal to `H`; however, for those frequencies at which $X(e^{j\omega})$ is essentially zero, set `H2` to zero.

(j). Plot the magnitude of the frequency response given by `H2`. Note that this frequency response magnitude varies rather smoothly with $\omega$, except where the frequency response was arbitrarily set to zero in Part ??. However, the frequency response $H(e^{j\omega})$ can be estimated at these frequencies by assuming that $H(e^{j\omega})$ is a smooth function. Under this assumption, the values of $H(e^{j\omega})$ at frequencies for which $X(e^{j\omega})$ is essentially zero can be interpolated from the values of $H(e^{j\omega})$ at neighboring frequencies.

(k). For any frequency $\omega_0$ for which $X(e^{j\omega_0})$ is assumed to be zero, set the value of $H(e^{j\omega_0})$ in `H2` by linearly interpolating from the values of $H(e^{j\omega})$ at the two frequencies $\omega_k$ nearest to $\omega_0$. Because $H(e^{j\omega})$ is a complex valued function, linearly interpolate the real and imaginary components of $H(e^{j\omega_0})$ separately. Plot the magnitude and phase of `H2`.

(l). Using `ifft`, compute from `H2` the values of $h[n]$ on the interval $0 \leq n \leq N-1$. Store this impulse response in the vector `h2`. Verify this impulse response by computing the convolution `y2=conv(h2,x)`, and comparing this output on the interval $0 \leq n \leq N-1$ with $y[n]$ in Eq (??). Does the assumption that $H(e^{j\omega})$ is smooth improve your result?

## ■ 5.6 Partial Fraction Expansion for Discrete-Time Systems

Discrete-time LTI systems are often modeled by linear constant-coefficient difference equations, whose frequency responses are a ratio of polynomials in $e^{-j\omega}$. When the ratio of polynomials is expressed as a partial fraction expansion, the impulse response can be determined by inspection or from a table of DTFT pairs. If the DTFT of the input to an LTI system satisfying a linear constant-coefficient difference equation is also a rational polynomial in $e^{-j\omega}$, then a partial fraction expansion can be used to calculate the corresponding output of the system. The function `residue` computes the coefficients of a partial fraction expansion, and is particularly useful when the denominator of the rational polynomial has high order, since high-order polynomials can be difficult to factor without the aid of a computer. In this example, you will derive the impulse responses of discrete-time LTI systems using `residue` and will then verify your results using `filter`.

The function `[r,p,k]=residue(b,a)` computes the coefficients of a partial fraction expansion

$$\frac{b_M\, z^M + \cdots + b_1\, z + b_0}{a_N\, z^N + \cdots + a_1\, z + a_0} = \sum_{n=1}^{N} \frac{r_n}{(z - p_n)^{d_n}} + \sum_{m=0}^{M-N} k_m\, z^m ,$$

where some of the roots $p_n$ may be repeated. The exponent $d_n$ is equal to one, unless $p_n$ is a repeated root. If the root $p_n$ is repeated $q$ times, i.e., $p_n = p_{n+1} = \cdots = p_{n+q-1}$, then $d_{n+i} = 1 + i$ for $i = 0, \ldots, q-1$. The vector `k` returned by `residue` contains the coefficients of the second sum, $k_m$, and is a null vector unless $M \geq N$. Note that because `residue` operates on polynomials in any arbitrary variable, called $z$ here for convenience, this function can be used for both continuous-time frequency responses where $z = j\omega$ and discrete-time frequency responses where $z = e^{-j\omega}$. In Exercise ??, `residue` is used to compute the partial fraction expansion of continuous-time frequency responses, $H(j\omega)$, and the exercise also covers how `residue` handles repeated roots $p_n$. For stable discrete-time LTI systems, the coefficients computed by `residue` will correspond to a partial fraction expansion of the form

$$H(e^{j\omega}) = \sum_{n=1}^{N} \frac{r_n}{(e^{-j\omega} - p_n)^{d_n}} + \sum_{m=0}^{M-N} k_m\, e^{-jm\omega} . \tag{5.14}$$

The inverse DTFT of each term in Eq. (??) can be computed by inspection or with the aid of a table of DTFT pairs.

## Basic Problems

Consider the discrete-time LTI system whose input and output satisfy

$$6\, y[n] - 5\, y[n-1] + y[n-2] = x[n] - x[n-1] .$$

(a). Create two vectors `a` and `b`, which contain the coefficients of the difference equation in the form required by `filter`.

(b). Analytically determine the frequency response, $H(e^{j\omega})$, of the system. Create two vectors `num` and `den`, which contain the coefficients of the numerator and denominator polynomials of the frequency response. The coefficients should be ordered in descending powers of $e^{-j\omega}$. For instance, the vector `[1 2 3]` contains the coefficients of the polynomial $3 + 2e^{-j\omega} + e^{-j2\omega}$. Note that this ordering is the reverse of the ordering required for `a` and `b` by `filter`.

(c). Use `residue(num,den)` to determine the partial fraction expansion of the frequency response. From this partial fraction expansion, determine an analytic expression for the impulse response.

(d). Use `filter(b,a,x)` to compute the impulse response of the system on the interval $0 \leq n \leq 10$, where `x` contains the unit impulse on the same interval, and use this response to verify the analytic expression determined in Part ??.

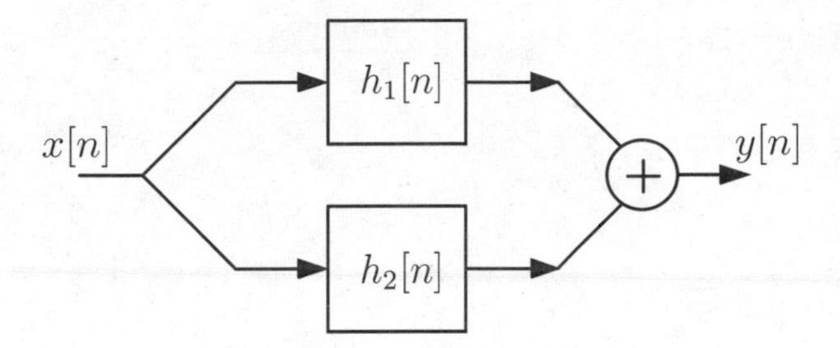


**Figure 5.2.** A parallel realization of the LTI system satisfying Eq. (??).

## Intermediate Problems

Consider the system illustrated in Figure ??. The input and output of this system are known to satisfy

$$36\,y[n] - 33\,y[n-1] + 10\,y[n-2] - y[n-3] = 39\,x[n-1] - 21\,x[n-2] + 3\,x[n-3]\,, \qquad (5.15)$$

and the impulse response of the system in the lower branch is $h_2[n] = 3\,(1/4)^n\,u[n]$. You will be asked to determine $h_1[n]$, the impulse response of the system in the upper branch.

(e). From Eq. (??), determine the frequency response $H(e^{j\omega})$ of the overall system.

(f). Using `residue`, determine the partial fraction expansion of $H(e^{j\omega})$. Use this partial fraction expansion to determine an analytical expression for $h[n]$, the impulse response of the system satisfying Eq. (??).

(g). Use `filter` to compute the impulse response of the system on the interval $0 \le n \le 32$, and use this response to verify the analytic expression you derived in Part ??.

(h). Determine $h_1[n]$, the impulse response of the system in the upper branch. What is the difference equation satisfied by the input and output of the system in the upper branch?

## Advanced Problem

Consider again the LTI system whose input and output satisfy Eq. (??).

(i). For the following three inputs, state whether or not `residue` can be used to compute $y[n]$:

$$x_1[n] = \left(\frac{1}{3}\right)^{|n-1|},$$

$$x_2[n] = \begin{cases} 1, & \text{for } 0 \le n \le 10\,, \\ 0, & \text{otherwise}\,, \end{cases}$$

$$x_3[n] = \frac{\sin(3\pi n/4)}{3\pi n/4}.$$

Explain your answers.

Chapter 6

# Time and Frequency Analysis of Signals and Systems

This chapter studies the time and frequency analysis of signals and systems, pulling together many of the concepts developed in the text so far. While studying either the time-domain or the frequency-domain characteristics of systems can be very useful, jointly understanding how these properties are related yields a more complete understanding of signals and linear time-invariant systems. This chapter applies many of the time and frequency analysis techniques that you have learned in previous chapters to both discrete- and continuous-time systems. For example, in Exercise ?? you will learn how the coefficients of a second-order differential equation describing an automobile shock absorber affect the behavior of the car suspension, as evidenced in both the frequency and time domains. This exercise will make use of the functions `freqs` and `lsim` for the frequency-domain and time-domain analyses. You will also explore relationships between the behavior of the system in both domains, such as the bandwidth and the reaction time of the shock absorber. These relationships are also examined in Exercise ?? while applying the filter design routines `butter` and `remez` to image processing. Exercise ?? shows how to manipulate the impulse response of a filter in the time domain to achieve a variety of different frequency responses. While many of the filtering exercises have dealt with the effects of the magnitude of the frequency response on the input signal, Exercise ?? demonstrates how the phase response of a discrete-time filter can alter the time characteristics of recorded speech. The function `unwrap` is used to help view the continuous phase response of the filter. Exercise ?? continues the discussion of the effect of phase on narrowband signals in the context of a frequency-division multiple-access (FDMA) communication system. In this exercise, the group delay of discrete-time filters is used to separate in time a set of signals which have narrow bandwidth. This exercise discusses some of the relationships between the group delay of a filter, calculated with the function `grpdelay`, and the time response of the filter to narrowband inputs. The last exercise in the chapter, Exercise ??, discusses the problem of filter design in the context of predicting the stock market. Contrary to the frequency-domain characterizations of most filter design problems, in this exercise the filter design is completely specified in the time domain in the form of a least-squares prediction problem which is solved using the MATLAB \ operator. Relationships between the modeling errors in the time domain and the frequency domain are also explored.

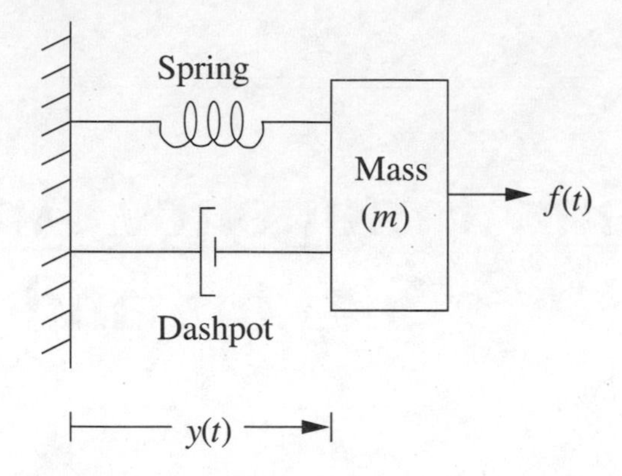


**Figure 6.1.** A spring-and-dashpot shock absorber.

## 6.1 A Second-Order Shock Absorber

The second-order linear constant-coefficient differential equation of the form

$$\frac{d^2y(t)}{dt^2} + 2\zeta\omega_n \frac{dy(t)}{dt} + {\omega_n}^2 y(t) = {\omega_n}^2 x(t) \tag{6.1}$$

can be used to model many physical systems. Two examples are RLC circuits and mechanical systems like the shock-absorber system illustrated in Figure ??, which consists of a spring and a viscous dashpot. The form of Eq. (??) is also very useful because $\omega_n$ and $\zeta$ are directly related to important properties of these systems. For reasons which will become clear, $\omega_n$ is referred to as the undamped natural frequency and $\zeta$ is referred to as the damping ratio.
For the shock-absorber system illustrated in Figure ??, the signal $f(t)$ is the external force applied to the mass $m$, and $y(t)$ is the horizontal displacement from the equilibrium point at which the spring exerts no restorative force. Throughout this exercise, the effects of gravity will be ignored. The restorative force is given by

$$F_{\text{spring}} = -ky\,,$$

while the force exerted by the dashpot is given by

$$F_{\text{dash}} = -v\frac{dy}{dt}\,.$$

The parameter $k$ is called the spring constant and $v$ is a measure of the viscosity of the dashpot. Summing the forces acting upon the mass connected to the shock absorber and for convenience substituting $f(t) = kx(t)$ for the external force, we obtain the following differential equation:

$$\frac{d^2y(t)}{dt^2} + \left(\frac{v}{m}\right)\frac{dy(t)}{dt} + \left(\frac{k}{m}\right)y(t) = \left(\frac{k}{m}\right)x(t)\,. \tag{6.2}$$

Comparing Eqs. (??) and (??), we can assign

$$\omega_n = \sqrt{\frac{k}{m}} \quad \text{and} \quad \zeta = \frac{v}{2\sqrt{km}}.$$

As you will see, the shock absorber acts as a lowpass filter on the forcing term $x(t)$, damping high-frequency fluctuations due to the road surface and allowing for a smoother ride. However, if the spring constant and the viscosity of the dashpot are not chosen carefully, the shock absorber may respond too slowly to bumps in the road surface, leading to a loss of control of the vehicle. In this exercise you will become familiar with how the values of $k$ and $v$, and hence the values of $\omega_n$ and $\zeta$, affect the performance of the shock absorber. Since shock absorbers are usually chosen with a specific vehicle mass in mind, assume for the rest of this exercise that $m = 1$. In this case, the parameters $\omega_n$ and $\zeta$ are determined completely by the shock-absorber coefficients $k$ and $v$.

## Basic Problems

(a). Analytically find the frequency response of the LTI system which satisfies Eq. (??). What is the value of $H(j\omega)$ at $\omega = 0$? To what value does $\log_{10}|H(j\omega)|/\log_{10}\omega$ converge for large values of $\omega$? These two values are known as the asymptotes of the Bode plot of $|H(j\omega)|$, and will be useful for verifying and analyzing the plots you will obtain.

In the following problems, you will fix $\zeta = 1/\sqrt{2}$ and determine how the shock-absorbing system changes with the value of $\omega_n$:

(b). For $k = 1$, determine the coefficients of the numerator and denominator polynomials of the frequency response and store these values in the vectors `b` and `a`. Use the function `freqs`, which is described in Tutorial ??, to determine $H(j\omega)$ at the frequencies contained in the vector `W=logspace(-2,2,100)`. Store these samples of $H(j\omega)$ in the matrix `H1`. Note that `W` contains evenly spaced samples of $\log_{10}\omega$ rather than of $\omega$. The reason for this sampling is that you will eventually plot $\log_{10}|H(j\omega)|$ versus $\log_{10}\omega$.

(c). For $k = 0.09$ and $k = 4$, determine $H(j\omega)$ at the frequency samples in `W` and store these samples in `H2` and `H3`, respectively.

(d). Use `loglog` to plot the magnitude of `H1`, `H2`, and `H3` versus `W` on the same set of axes. The cutoff frequency of the shock-absorbing system can be loosely defined as the "knee" of the log-log plot of $|H(j\omega)|$. For frequencies higher than the cutoff frequency, the frequency response magnitude decreases rapidly with increasing $\omega$. Based upon your plots, is $\omega_n$ roughly equal to the cutoff frequency of each shock absorber? Do the asymptotes of your plots agree with the values determined in Part ???

(e). For the three shock-absorbing systems given by $k = 0.09$, 1, and 4, use `step` to determine the step response of each system at the time samples contained in the vector `t=linspace(0,30)`. Plot the three step responses on a single set of axes. How does the rise time, i.e., the time for the step response to first reach its steady-state value, change as a function of $\omega_n$? How is the rise time related to the bandwidth of the filter, i.e., the cutoff frequency? Would you expect a shock absorber with small bandwidth to react quickly or slowly to rapid changes in the external force $x(t)$?

## Intermediate Problems

You will now consider how the characteristics of the shock-absorbing system change with the value of the damping ratio $\zeta$ for the fixed value of $\omega_n = 1$.

(f). For the frequency samples in `W`, use `freqs` to determine the values of the frequency response for the shock-absorbing systems given by $v = 2$, $\sqrt{2}/2$, $1/2$, and $1/3$. Store these frequency response samples in the vectors `H4`, `H5`, `H6`, and `H7`.

(g). Use `loglog` to plot on a single set of axes the magnitude of the frequency responses `H4` through `H7`. How does the value of $|H(j\omega_n)|$, the magnitude of the frequency response at the natural frequency, change with the viscosity $v$? Provide a physical explanation for this relationship. Does the value of $v$ significantly affect the frequency response at frequencies far from $\omega_n$?

(h). For each of the four shock-absorbing systems, use `step` to determine the step response of each system at the time samples in `t=linspace(0,30)`. Plot the four step responses on a single set of axes. How does the rise time vary with the viscosity $v$? How do the oscillations and overshoot in the step response vary with $v$? (Overshoot occurs whenever the step response exceeds its steady-state value.)

(i). Approximately determine the value of $v$ for which the step response rises the fastest but has no overshoot. Plot this step response, and note the corresponding value of $\zeta$.

While the spring constant $k$ controls the trade-off between the bandwidth and the rise time (response time) of the shock-absorbing system, the viscosity can be decreased in order to decrease the rise time without significantly changing the cutoff frequency. The price, however, is that as $v$ decreases and the rise time decreases, the system experiences oscillations in the time domain. These oscillations are manifested in the frequency response by an increase in the peak value of $|H(j\omega)|$ near $\omega_n$. Therefore $k$ and $v$ must be chosen with both the frequency-domain and the time-domain characteristics of the shock-absorbing system in mind.

## Advanced Problems

An important property of any automobile suspension system is its ability to cushion its passengers from rapid variations in the road surface. An automobile suspension system using the spring-dashpot shock absorber is illustrated in Figure ??. For this system, the height of the road is defined to be $x(t)$ and center of mass of the automobile is $z(t) = y(t) + y_0$. The displacement of the spring is given by

$$\begin{aligned} d(t) &= z(t) - x(t) - y_0\,, \\ &= y(t) - x(t)\,, \end{aligned}$$

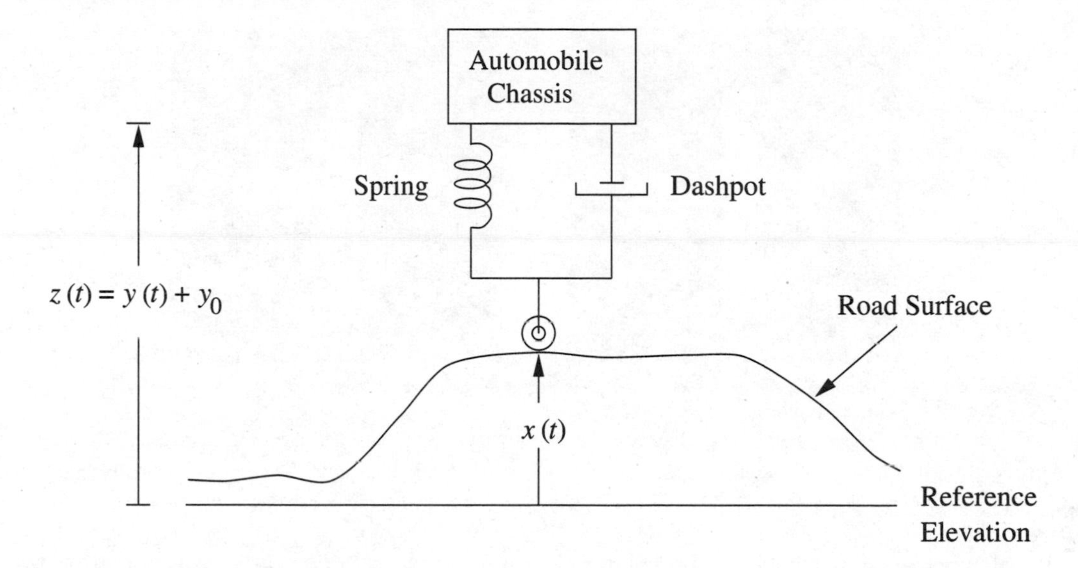


**Figure 6.2.** A model of an automobile suspension system consisting of a spring and a dashpot.

so that the spring exerts no force when $z(t) - x(t) = y_0$. The automobile suspension system satisfies the following differential equation:

$$\frac{d^2y(t)}{dt^2} + \left(\frac{v}{m}\right)\frac{dy(t)}{dt} + \left(\frac{k}{m}\right)y(t) = \left(\frac{v}{m}\right)\frac{dx(t)}{dt} + \left(\frac{k}{m}\right)x(t)\,. \tag{6.3}$$

Note the difference between the spring-dashpot system illustrated in Figure ?? and the system in Figure ??. While the external force is exerted directly upon the mass in Figure ??, in Figure ?? the force is exerted indirectly (on the other end of the spring and dashpot) by the road surface.

In the following problems you will use the intuition you developed in the previous problems to select the values of $k$ and $v$ so that the corresponding suspension system responds desirably to a given road surface. Assume for this problem that $m = 1$ and $y_0 = 0.5$. In addition to smoothing the ride of the passengers, you will also have to consider additional constraints upon the suspension system. The system must obey

$$z(t) > x(t) \qquad \text{and} \qquad |y(t) - x(t)| < y_1\,. \tag{6.4}$$

The first constraint ensures that the automobile chassis does not crash into the ground, and the second constraint reflects the limited range of displacement of any shock absorber. For this exercise, assume that $y_1 = 0.5$.

(j). For these problems, you will need to load the file `road.mat`. This file is in the Computer Explorations Toolbox, which is available from The MathWorks at the address listed in the Preface. If `road.mat` is in your MATLABPATH, then typing `load road` should load the vectors `x` and `t` into your workspace. The vector `x` contains the height of a "typical" road surface at the time samples in `t`. The signal $x(t)$ consists of a

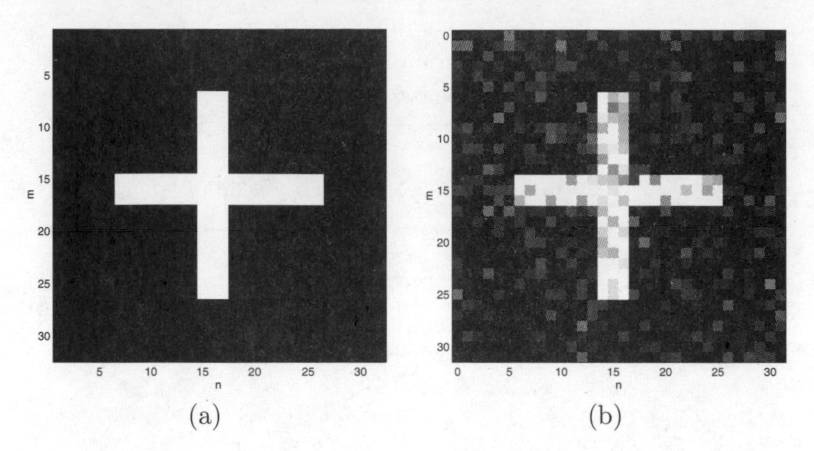


**Figure 6.3.** Two 32-by-32 pixel images: (a) the "plus sign", and (b) a noisy version of the "plus sign".

slowly varying surface in the form of a trapezoid, plus fine-scale fluctuations due to the road surface. Determine a pair of values for $k$ and $v$ such that (1) the fine-scale fluctuations of the road surface are smoothed, and (2) the mechanical constraints are obeyed. Use `lsim` to compute the output signal $y(t)$. Plot $z(t)$ and $x(t)$ on the same pair of axes. Note that there are a range of values for $k$ and $v$ for which the suspension satisfies the constraints of Eq. (??). You might plot $z(t)$ for two pairs of values for $k$ and $v$ and then discuss the relative merits of the two systems.

(k). While the suspension system is usually designed with a specific mass $m$ in mind, the vehicle mass will change as the number of passengers and cargo weight changes. For your chosen values of $k$ and $v$, what is the range of values for $m$ such that the system still obeys the constraints of Eq. (??)?

## 6.2 Image Processing with One-Dimensional Filters

A two-dimensional (2D) discrete-time signal $x[m, n]$ has two index variables rather than one. An image is a 2D discrete-time signal which is nonzero only over a finite region, e.g., the $N$-by-$N$ square $0 \leq m, n \leq N-1$. For such images, $m$ corresponds to the row index while $n$ corresponds to the column index. Two 32-by-32 pixel images are illustrated in Figure ??. (A pixel is the name for a sample of an image, and is short for *picture element*.) Many of the time-domain and frequency-domain tools used for designing and analyzing one-dimensional (1D) signals apply equally well to 2D signals.[1] In particular, a widely-used method for filtering 2D signals is to process the signal with a cascade of two 1D filters, one filter for each direction. The cascade of these two 1D filters, called a separable 2D filter, is given by the following two steps: (i) for each $n$ compute the output $z[m, n]$ of the LTI

[1]For further discussion of two-dimensional signals and systems, see *Multidimensional Digital Signal Processing* by Dudgeon and Mersereau, 1984, or *Two-Dimensional Signal and Image Processing* by Lim, 1990.

system which satisfies

$$\sum_{k=-K}^{K} a_k z[m-k,n] = \sum_{j=-K}^{K} b_j x[m-j,n]\,; \tag{6.5}$$

(ii) for each $m$ compute the output $y[m,n]$ of the LTI system which satisfies

$$\sum_{k=-K}^{K} c_k y[m,n-k] = \sum_{j=-K}^{K} d_j z[m,n-j]\,. \tag{6.6}$$

The output $y[m,n]$ is the filtered version of $x[m,n]$.
Note that both the column-wise filter defined by the coefficients $a_k$ and $b_j$ and the row-wise filter defined by the coefficients $c_k$ and $d_j$ are noncausal. In the context of image processing, causality is not a constraint, since one is often given the entire image before it is to be processed.
In this exercise, you will lowpass filter an image using a cascade of two 1D filters. For simplicity, the two cascaded 1D filters are identical, i.e., $c_k = a_k$ and $d_j = b_j$. You will see that the time-domain and frequency-domain tools you have learned for 1D signals can provide considerable insight into 2D filtering.
For this exercise, you will need to load the data file `plus.mat`, which is provided in the Computer Explorations Toolbox. If this file is already in your MATLABPATH, you can load the data by typing `load plus`. If loaded correctly, your workspace will then contain a matrix `x`, which contains the image to be processed in this exercise, and a noise-corrupted version of `x` in the matrix `xn`, which will be filtered in the Advanced Problems. To display the image, type

```
>> colormap(gray);
>> image(64*x);
```

Your image should be identical to the one in Figure ??(a). The command `colormap(gray)` is used to select the mapping from the values of `x` to the colors displayed by `image`. Also, because the colormap has 64 color levels, the input to `image` is scale by 64.

## Basic Problems

In these problems you will create three 1D lowpass filters, each of which will eventually be used to lowpass filter the image contained in `x`. Analyzing the properties of these filters in 1D will be useful for interpreting the images filtered by a cascade of two 1D filters.

(a). The functions `butter` and `remez` both return the coefficients of difference equations which define 1D discrete-time filters. The filters determined by `butter` have infinite-length impulse responses, while the filters determined by `remez` have finite-length impulse responses. Type

```
>> wc = 0.4;
>> n1 = 10; n2 = 4; n3 = 12;
```

```
>> [b1,a1] = butter(n1,wc);
>> a2 = 1; b2 = remez(n2,[0 wc-0.04 wc+0.04 1],[1 1 0 0]);
>> a3 = 1; b3 = remez(n3,[0 wc-0.04 wc+0.04 1],[1 1 0 0]);
```

where `wc` is the cutoff frequency of each filter (normalized by $\pi$) and `n1`, `n2`, and `n3` are the orders of the three filters.

(b). Use `freqz`, which is described in Tutorial ??, to plot the magnitude and phase of each of the three 1D filters determined in Part ??. As a check, verify that `wc` is the approximate cutoff frequency of each filter. Which filters have linear phase?

(c). Use `filter` to determine the step response of each filter for `n=[0:20]`. Plot each of these step responses. Which step response has the largest overshoot, i.e., the difference between the steady-state value and the maximum value of the step response? You will see that large overshoots or ringing in the step response can lead to undesirable artifacts in the filtered images.

## Intermediate Problems

For image processing applications, noncausal filters can be used and are in fact desired. One reason for using noncausal filters is that the contents of an image should not be shifted in location after processing. In the following problems, you will first learn how to implement 1D noncausal filters using `filter`. Then you will write an M-file which will repeatedly call `filter` to process the columns and rows of the image stored in `x`.
The function `filter` assumes that the filter coefficients stored in the vectors `a` and `b` correspond to a causal filter. For example, if `x16` contains the column of the image for $n = 16$, i.e., $x[m, 16]$ for $0 \leq m \leq N-1$, then `y16=filter(b,a,x16)` stores in `y16` the output of the causal filter

$$\sum_{k=0}^{2*\mathtt{d}} \mathtt{a}(k+1)\, y[m-k, 16] = \sum_{j=0}^{2*\mathtt{d}} \mathtt{b}(j+1)\, x[m-j, 16] \qquad (6.7)$$

on the interval $0 \leq m \leq N-1$, where the vectors `a` and `b` have `2*d+1` elements. For this exercise, however, you will need to implement the following noncausal filter using `filter`:

$$\sum_{k=-\mathtt{d}}^{\mathtt{d}} \mathtt{a}(k+1+\mathtt{d})\, y[m-k, 16] = \sum_{j=-\mathtt{d}}^{\mathtt{d}} \mathtt{b}(j+1+\mathtt{d})\, x[m-j, 16]\,. \qquad (6.8)$$

This filter is in the same form as the noncausal filters given by Eqs. (??) and (??). Noting the similarity between Eqs. (??) and (??), you can see that `y16=filter(b,a,x16)` computes the output of the noncausal filter on the interval $-\mathtt{d} \leq m \leq N-\mathtt{d}-1$. In other words, `d` is equal to the advance induced by `filter`. If the vector returned by `filter` is to contain the output of the noncausal filter for $n = 16$ and $0 \leq m \leq N-1$, the input to `filter` must contain $x[m, 16]$ for $0 \leq m \leq N+\mathtt{d}-1$. Since $x[m, 16]$ is not known outside the interval $0 \leq m \leq N-1$, the values outside this interval can assumed to be zero. Therefore, the output of the noncausal filter on the interval $0 \leq m \leq N-1$ can be extracted from the last $N$ samples the signal produced by `filter(b,a,[x16; zeros(d,1)]`.

(d). Create the column vector `x16=x(:,16)`, which provides a 1D signal defined on the interval $0 \leq m \leq 31$. Use `filter` to compute the output of the noncausal filter given by Eq. (??) for `a=a1` and `b=b1`, assuming the input is given by `x16` for $0 \leq m \leq 31$ and is zero elsewhere. Store in `y16` the filter response on the interval $0 \leq m \leq 31$. Plot `x16` versus `y16`. If `y16` is computed correctly, the discontinuities in `x16` line up with the "smoothed" discontinuities in `y16`.

(e). For filters two and three, determine the response of these 1D filters to the column vector `x16`. For each filter, you will have to append zeros to `x16` before calling `filter`. Again plot the responses versus `x16` to ensure that the discontinuities are aligned and there is no delay or advance in your filter implementation.

(f). Write an M-file for the function `filt2d` which noncausally filters the 2D image stored in the matrix `x`. The first line of the your M-file `filt2d.m` should read

```
function y = filt2d(b,a,d,x)
```

The vectors `a` and `b` contain the coefficients of the 1D filter and `d` contains the "delay" associated with a causal implementation of the filter. (Note that `d` is just `n/2`, where `n` is the order of the 1D filter). Your M-file should essentially consist of two steps: (i) filtering each column of an $N$-by-$N$ matrix `x` and storing the results in an $N$-by-$N$ matrix `z`, followed by (ii) filtering each row of `z`. Note that you will have to append an appropriate number of zeros to each column of `x` and each row of `z` before filtering. The matrix `y` returned by `filt2d` should have the same dimensions as `x`.

(g). Use `image` to display the filtered image given by `filt2d` for each of the three filters. (Remember to scale the input to `image` by 64.) Make sure that the image has not been delayed, i.e., the filtered "plus" contained in `y` should have the same center location as the "plus" in `x`.

(h). Which filter leads to more distortion in the shape of the original image? Pay attention to any destruction of the symmetries present in the input image. This distortion is due in part to the nonlinearity of the phase in one of the filters.

## Advanced Problems

Now you will compare the outputs of the second and third filters. These filters both have finite-length impulse responses but require different numbers of coefficients.

(i). Store in `y2` and `y3` the matrices returned by filtering `x` with `filt2d` using the second and third 1D filters.

(j). Compare `y2` and `y3` using `image`. Which output image has more oscillations? How could you have predicted the oscillations from the step responses of the two filters?

For many image processing applications, filters with smooth transition bands are preferred over filters which more closely approximate the frequency response of an ideal lowpass filter, which has a sharp transition band. One application which illustrates the advantages of a smoother transition band is the removal of noise from images.

(k). Display the noisy image using `image(64*xn)`. Your image should be identical to the one in Figure ??(b).

(l). Determine the output of `filt2d` for the second and third filters for the noisy input `xn`. Display both outputs using `image`. Which filter does a better job of removing the noise? Pay particularly close attention to the output signals in areas for which `x` is zero. You may wish to explore the effect of changing the parameters of the filters returned by `butter` and `remez`.

## 6.3 Filter Design by Transformation

A common technique used to design frequency-selective filters is to start with a prototype lowpass filter and transform that filter's impulse response to obtain a filter with the desired frequency response. In this exercise, you will learn how to manipulate a finite-length impulse response (FIR) lowpass filter to get a highpass filter, a bandpass filter, a bandstop filter, and several multi-band filters. The prototype lowpass filter you will be working with is an approximation to the ideal lowpass filter with a cutoff frequency of $\pi/5$. The frequency response of the ideal filter is

$$H_{\text{id}}(e^{j\omega}) = \begin{cases} 1, & |\omega| < \pi/5, \\ 0, & \pi/5 \leq |\omega| \leq \pi. \end{cases} \tag{6.9}$$

The impulse response for the prototype FIR filter whose frequency response magnitude approximates $H_{\text{id}}(e^{j\omega})$ is in the data file `protoh.mat`. This file is provided in the Computer Explorations Toolbox. If this file is already in a directory in your MATLABPATH, you should be able to load it into MATLAB by typing `load protoh`. The variable `h` in the MATLAB workspace will be the impulse response of the prototype filter you will use as the basis for designing other filters in this exercise.

### Basic Problems

(a). For the filter with impulse response `h`, compute the frequency response at 1024 evenly spaced samples in $\omega$ using `freqz` and store these values in `H`. Plot the magnitude of `H` and confirm that it approximates the desired frequency response magnitude $|H_{\text{id}}(e^{j\omega})|$. Also, plot the impulse response `h` using `stem`. Analytically calculate the impulse response of the ideal lowpass filter specified by Eq. (??). Is `h` similar in form to the impulse response of the ideal lowpass filter?

(b). The modulation property for the DTFT can be used to construct other filters from a lowpass filter. The modulation property states that if $X(e^{j\omega})$ is the DTFT of $x[n]$, then the DTFT of $e^{j\omega_0 n}x[n]$ is $X(e^{j(\omega-\omega_0)})$. Use `h` and this property to design a real-valued impulse response $h_1[n]$ for a highpass filter whose frequency response magnitude approximates the ideal highpass filter:

$$H_{\text{hpf}}(e^{j\omega}) = \begin{cases} 1, & 4\pi/5 < |\omega| \leq \pi, \\ 0, & \text{otherwise}. \end{cases} \tag{6.10}$$

Store the impulse response of your new filter in `h1`. Generate an appropriately labeled plot of the magnitude of the frequency response for your highpass filter. Also, plot `h1` using `stem`. How are `h` and `h1` related?

(c). Use `h` and the modulation property to design a bandpass filter whose frequency response approximates an ideal bandpass filter with the following frequency response between $-\pi$ and $\pi$:

$$H_{\mathrm{bpf}}(e^{j\omega}) = \begin{cases} 1, & 3\pi/10 \le |\omega| \le 7\pi/10\,, \\ 0, & \text{otherwise}\,. \end{cases} \tag{6.11}$$

Store the real-valued impulse response of your bandpass filter in `h2`. Generate an appropriately labeled plot demonstrating the magnitude of the frequency response of your filter approximates the ideal bandpass filter in Eq. (??). Also, plot `h2` using `stem`. How are `h` and `h2` related?

(d). Use linearity and the modulation property to design a bandstop filter from `h` whose frequency response approximates an ideal bandstop filter with the following frequency response between $-\pi$ and $\pi$:

$$H_{\mathrm{bsf}}(e^{j\omega}) = \begin{cases} 0, & 3\pi/10 \le |\omega| \le 7\pi/10\,, \\ 1, & \text{otherwise}\,. \end{cases} \tag{6.12}$$

Store the real-valued impulse response of your bandstop filter in `h3`. Generate an appropriately labeled plot demonstrating that the magnitude of the frequency response of your filter approximates the ideal bandstop filter above.

## Intermediate Problems

Sample rate expansion and compression are also useful for modifying the impulse response of a lowpass filter to get other filters. Sample rate expansion by an integer factor $L$ is defined as the operation

$$h_{\mathrm{e}}[n] = \begin{cases} h[n/L]\,, & n = kL \text{ and } k \text{ integer}\,, \\ 0\,, & \text{otherwise}\,, \end{cases} \tag{6.13}$$

while sample rate compression by an integer factor $M$ is defined as

$$h_{\mathrm{c}}[n] = h[Mn]. \tag{6.14}$$

It can be shown that expansion scales the frequency response such that $H_{\mathrm{e}}(e^{j\omega}) = H(e^{jL\omega})$. Pictorially speaking, the frequency response is compressed in $\omega$ by a factor of $L$, and is therefore periodic in $\omega$ with period $2\pi/L$. Computing the effect of sample rate compression of the impulse response on the frequency response is generally more complicated. The frequency response of the compressed filter is

$$H_{\mathrm{c}}(e^{j\omega}) = \frac{1}{M}\sum_{k=0}^{M-1} H\left(e^{j(\omega/M - 2\pi k/M)}\right). \tag{6.15}$$

Pictorially, this equation puts copies of $(1/M)H(e^{j\omega})$ centered at $\omega = 2\pi k/M$, then spreads the frequency axis by a factor of $M$ so that $H_c(e^{j\omega})$ remains periodic with period $2\pi$. If the original frequency response $H(e^{j\omega}) = 0$ for $\pi/M \le |\omega| \le \pi$, then sample rate compression can be thought of as making the original $H(e^{j\omega})$ wider by a factor of $M$ and scaled in amplitude by a factor of $1/M$. When this condition is not met, the copies of $H(e^{j\omega})$ may overlap and distort each other, which is known as aliasing. Chapter ?? explores this issue in more detail.

(e). Use the sample rate compression property to design a new lowpass filter from `h` which approximates

$$H(e^{j\omega}) = \begin{cases} 1, & |\omega| < 2\pi/5, \\ 0, & \text{otherwise}. \end{cases} \tag{6.16}$$

Store the real-valued impulse response of this new filter in `h4` and generate an appropriately labeled plot of the magnitude of the frequency response of `h4` to confirm that it approximates the desired frequency response.

(f). Use the sample rate expansion property to design a new multiple bandpass filter which approximates the following frequency response between $-\pi$ and $\pi$:

$$H(e^{j\omega}) = \begin{cases} 1, & |\omega| < \pi/15, \\ 1, & 9\pi/15 < |\omega| < 11\pi/15, \\ 0, & \text{otherwise}. \end{cases} \tag{6.17}$$

Store the real-valued impulse response of this new filter in `h5` and generate an appropriately labeled plot of the magnitude of the frequency response of `h5` to confirm that it approximates the desired frequency response.

## Advanced Problems

Combinations of two or more of the modulation property, sample rate expansion property, and sample rate compression property allow you to design a wide variety of filters. For each of the following ideal filters, use these properties and `h` to find the real-valued impulse response of a new filter approximating the ideal frequency response given. Generate a plot of the magnitude of the frequency response of the filter with the impulse response you specify.

(g).

$$H(e^{j\omega}) = \begin{cases} 1, & |\omega| < \pi/5, \\ 2, & \pi/5 \le |\omega| \le 3\pi/5, \\ 0, & \text{otherwise}. \end{cases} \tag{6.18}$$

(h).

$$H(e^{j\omega}) = \begin{cases} 1, & |\omega| < 2\pi/15, \\ 1, & 8\pi/15 < |\omega| < 12\pi/15, \\ 0, & \text{otherwise}. \end{cases} \tag{6.19}$$

## 6.4 Phase Effects for Lowpass Filters

Filters are often designed to meet a specification of the magnitude of the frequency response. In this exercise, you will see that in some situations the effect of the phase can be important as well. You will examine the effect of processing clean speech and noisy speech with several filters which have roughly the same frequency response magnitude, and compare how the outputs of the different filters sound.

These problems require you to work with three different lowpass filters which are specified by coefficient vectors stored in the file `phdist.mat`, which is in the Computer Explorations Toolbox. If this file is already in your MATLABPATH, you can type `load phdist`. If this file has been loaded correctly, then typing `who` should result in

```
>> who
Your variables are:
a1          a3          b2          xnoise
a2          b1          b3
```

The vectors `a1` through `a3` and `b1` through `b3` contain the coefficient vectors for three discrete-time filters in the format required by `freqz` and `filter`. The vector `xnoise` contains a speech signal with some high-frequency noise added. You will also want to load the original version of this speech signal, which is the file `orig.mat` in the Computer Explorations Toolbox. If this file is in your MATLABPATH, you can load it by typing `load orig`. The uncorrupted version of this signal should now be in the vector `x`.

### Basic Problems

(a). Use `freqz` to compute the frequency response of all three filters at 1024 evenly spaced points on the interval $0 \leq \omega < \pi$. Plot the frequency response magnitude of all three filters. Suppose you used the same speech signal as the input to all three filters. Based on the magnitudes of the frequency responses, which two of the output signals would you expect to sound the most similar?

(b). Generate appropriately labeled plots of the phase of the frequency response for all three systems. The function `angle` will only return angles between $-\pi$ and $\pi$. If you just plot the output of `angle`, you may find it hard to understand some of the plots, since the phase will "wrap-around" from the top to the bottom of the graph every time it passes $\pi$ or $-\pi$. The function `unwrap` takes the output of `angle` and removes the "wrap-around" effect to obtain a more continuous phase function. For example, if `H` contains the samples of the frequency response at the frequencies specified in `omega`, then `plot(omega/pi,unwrap(angle(H)))` displays the unwrapped phase. Which two filters are the most alike in phase response?

(c). Define the vectors `y1` through `y3` to be the output of the filters when they are applied to the clean speech signal in `x` using `filter`. Play each speech signal using `soundsc`. Is the speech still easy to understand in each case? Which two speech signals sound most alike? Based on your results for this input signal, would you say that the magnitude or phase of these filters is more important?

### Intermediate Problems

In these problems, you will use the three filters to remove high-frequency noise from a corrupted version of the input signal. This is a simplified example of how the Dolby$^{\text{TM}}$ noise-reduction system works. Most tape decks add high-frequency noise to the signal they are recording. The Dolby system reduces this noise.

(d). Play both `x` and `xnoise` using `sound` to hear the difference between the clean and corrupted speech signals.

(e). Use `fft` to compute the DTFTs of `x` and `xnoise` at 8192 evenly spaced samples in $\omega$ and store these values in `X` and `Xn`. Plot the magnitude of the DTFT for each signal. Can you see the high-frequency noise in the DTFT of the corrupted signal?

(f). Let `yn1` through `yn3` be the result of filtering `xnoise` with each of the three filters. Listen to all three of these signals. Which filter best eliminated the noise?

(g). Use `fft` to compute the DTFT of the filtered speech signals at 8192 evenly spaced points in $\omega$ and store them in `Yn1` through `Yn3`. Make appropriately labeled plots of the magnitude of the DTFT for each signal. Do these plots show the differences you heard in Part **??**? Does the phase of the system used to remove the noise from the signal affect the intelligibility of the speech?

## 6.5 Frequency Division Multiple-Access

A standard method for sharing a channel among several users is through the use of frequency division multiple access, or FDMA. In a typical discrete-time FDMA system, a number of message values, $a_k$, are multiplied by a narrowband pulse, $p[n]$, and these pulses are each then modulated up to different frequencies by sinusoidal carrier signals of the form $\cos(\omega_k n)$. If the pulse is bandlimited and each of the signals $x_k[n] = a_k p[n]$ is modulated to disjoint frequency bands, then the message values can be uniquely recovered with bandpass filtering and demodulation. To be more precise, define the composite FDMA signal as

$$s[n] = \sum_{k=1}^{N} x_k[n]\cos(\omega_k n). \tag{6.20}$$

Assume the pulse $p[n]$ has a bandwidth of $\omega_x$, i.e., $X_k(e^{j\omega}) = 0$ for $|\omega| > \omega_x/2$. By also assuming that the signals are modulated to disjoint frequency bands, i.e., $|\omega_k - \omega_{k+1}| > \omega_x$, each of the signals $x_k[n]$ can be completely recovered from the signal $s[n]$ by frequency-selective filtering.

Rather than using frequency-selective filters to extract a single waveform, $x_k[n]$, in this exercise you will explore the use of the phase of all-pass filters to extract FDMA signals. Consider the response of a filter with impulse response $h[n]$ to one of the narrowband message signals $x_k[n]\cos(\omega_k n)$. The Fourier transform of the modulated signal is given by

$$\frac{1}{2}X_k\left(e^{j(\omega-\omega_k)}\right) + \frac{1}{2}X_k\left(e^{j(\omega+\omega_k)}\right) \tag{6.21}$$

and is zero outside of the small range of frequencies, $\omega_k - \omega_x/2 < |\omega| < \omega_k + \omega_x/2$. Therefore the effect of the filter upon this signal is limited to the behavior of $H(e^{j\omega})$ in the vicinity of $\omega_k$. For sufficiently small $\omega_x$, the phase response of the filter for $\omega$ near $\omega_k$ can be accurately modeled by a linear approximation:

$$\angle H(e^{j\omega}) \approx -\phi - \omega\alpha\,, \qquad \omega \text{ near } \omega_k. \tag{6.22}$$

For this approximation, the effect of the filter on the narrowband signal $x_k[n]\cos(\omega_k n)$ consists of an amplitude weighting by $|H(e^{j\omega_k})|$, an overall complex factor $e^{-j\phi}$, and a linear phase term, $e^{-j\omega\alpha}$ corresponding to a time delay of $\alpha$ samples. This time delay is referred to as the group delay at $\omega = \omega_k$, since it is the effective common delay experienced by the group of frequencies centered about $\omega_k$. The group delay at each frequency is the negative slope of the phase of $H(e^{j\omega})$ at that frequency and can be expressed as[2]

$$\tau(\omega) = -\frac{d}{d\omega}\left\{\angle H(e^{j\omega})\right\}, \tag{6.23}$$

so $\tau(\omega_k) = \alpha$ follows from Eq. (??).
For simplicity, you will assume that there are three users corresponding to the discrete-time frequencies $\omega_k = k\pi/4$, $k = 1, 2, 3$. Therefore the received signal will be of the form

$$s[n] = \sum_{k=1}^{3} a_k p[n]\cos(k\pi n/4). \tag{6.24}$$

You will also assume that $p[n]$ is the 75-sample Hamming window created by `p=hamming(75)`. The goal of this exercise is to use the group delay of an all-pass filter to extract each signal $x_k[n]$ from the transmitted signal $s[n]$. Specifically, you will use the all-pass system which satisfies

$$\begin{aligned} y[n] - 1.1172\,y[n-1] + 0.9841\,y[n-2] - 0.4022\,y[n-3] + 0.2247\,y[n-4] = \\ 0.2247\,x[n] - 0.4022\,x[n-1] + 0.9841\,x[n-2] - 1.1172\,x[n-3] + x[n-4]. \end{aligned} \tag{6.25}$$

## Basic Problems

(a). Create the vectors `a` and `b` for the all-pass filter. Store the frequency response of the filter in the vector `H` using `H=freqz(b,a,N,'whole')` for `N=1024`. Plot the frequency response magnitude to verify that the system is indeed all-pass.

(b). The function `[tau,w]=grpdelay(b,a,N,'whole')` returns in the vector `tau` the group delay of a discrete-time filter at `N` equally spaced samples of the frequency axis from 0 to $2\pi$. These frequency samples are returned in `w`. Use `grpdelay` to compute the group delay for the filter defined by the vectors `a` and `b`. Plot the group delay versus frequency.

[2] To be precise, group delay is defined in terms of the continuous phase to avoid evaluating the derivative at discontinuities.

(c). Using the pulse created by `p=hamming(75)`, construct the signals $s_1[n] = p[n]\cos(\omega_1 n)$, $s_2[n] = p[n]\cos(\omega_2 n)$, and $s_3[n] = p[n]\cos(\omega_3 n)$ and store them in the vectors `s1`, `s2`, and `s3`, i.e., assume that $a_k = 1$. Make each of these signals of length 512 by appending 437 zeros to each modulated pulse. Use `fft` to compute samples of the DTFT of each of these signals and store them in the vectors `S1`, `S2`, and `S3`. Plot the magnitudes of the DTFTs for each of the signals `s1`, `s2`, and `s3` on the same axes used to plot the group delay for the filter.

(d). Based on your plots from Part ??, determine the group delay of the filter over the range of each of the frequencies of the narrowband signals `s1`, `s2`, and `s3`. Since the filter is an all-pass filter, the DTFT magnitude of the output will be the same as the input. Also, since the signals $x_k[n]\cos(\omega_k n)$ are narrowband, the group delay of the filter is approximately constant over the range of frequencies for which the signals have appreciable energy. This implies that the envelope of the narrowband signal is simply delayed by the filter. Estimate the resulting delay that would be exhibited if each of the signals $x_k[n]\cos(\omega_k n)$ were processed by the filter. When $n_{d1}$ is an integer, you would expect that the envelope of $y1[n]$ would be approximately given by $s_1[n - n_{d1}]$, where $n_{d1}$ is the group delay of the filter at $\omega_1$, $\tau(\omega_1) = n_{d1}$. Compute and plot the response of the filter to each of the signals $s_k[n]$ using `filter` and compare the results with your estimates.

## Intermediate Problem

(e). Construct the signal `s=s1+s2+s3`, again assuming $a_k = 1$. Since each pulse $x_k[n]$ is 75 samples long, and the group delay in the filter is significantly less than 75 for all frequencies, processing the signal `s` with the filter will not be sufficient to separate the individual pulses. However, by repeatedly applying the filter to the signal, the associated group delays incurred by each pulse will accumulate, i.e., the envelope of the response to the $\ell$-th application of the filter, $s^{(\ell)}[n]$, will be approximately given by

$$x_1[n - \ell n_{d1}] + x_2[n - \ell n_{d2}] + x_3[n - \ell n_{d3}], \qquad (6.26)$$

where $n_{dk}$ is the group delay at $\omega_k$. Determine an appropriate number of applications of the filter, $\ell$, required to separate in time the pulses `s1`, `s2`, and `s3`. Repeatedly apply the filter $\ell$ times, and verify that the pulses do separate as you would expect. Plot $s^{(\ell)}[n]$ and indicate on the plot the value of $\ell n_{dk}$ for each $k$.

## Advanced Problems

(f). In addition to inducing a delay in the envelope of a narrowband pulse, the phase of the filter also affects the phase of the carrier signal, $\cos(\omega_k n)$. When the phase of the filter can be approximated as

$$\angle H(e^{j\omega})|_{\omega \approx \omega_k} \approx -\phi - \omega n_{dk}, \qquad \omega > 0,$$

verify analytically that the response of the filter to a narrowband pulse $x_k[n]\cos(\omega_k n)$ will be approximately given by $x_k[n - n_{dk}]\cos(\omega_k n - \phi - \omega_k n_{dk})$. For each of the frequencies $\omega_k$ determine the values of $\phi$ and $n_{dk}$ corresponding to the phase lag and group delay at each frequency.

(g). For $k = 1, 2, 3$, construct the signals $\hat{y}_k[n] = x_k[n-n_{dk}]\cos(\omega_k n - \phi - \omega_k n_{dk})$ based on your answers to Part ??. Compare each signal $\hat{y}_k[n]$ with the corresponding output of the filter produced using `filter` with $s_k[n]$ as input.

## 6.6 Linear Prediction on the Stock Market

Linear prediction is one of the most widely-used approaches to time series analysis involving applications such as speech coding, seismology, and frequency response modeling. In this exercise, you will learn how linear prediction can be used to design a discrete-time finite-length impulse response (FIR) filter to solve both a time-domain prediction problem and a frequency-domain modeling problem.

In the prediction problem, you observe a signal $x[n]$ and wish to design a system that can predict future values of the signal based solely upon past values. For linear prediction, this system is an FIR filter which computes a prediction based upon a linear combination of past values,

$$\hat{x}[n] = -\sum_{k=1}^{p} a_k x[n-k], \tag{6.27}$$

where $\hat{x}[n]$ is the predicted value of $x[n]$. Since $p$ previous values of the signal are used to formulate the prediction, this is a $p$th-order predictor. Given a fixed filter order, $p$, the linear prediction problem is to determine a set of filter coefficients, $a_k$, that best perform the prediction in Eq. (??). The most common measure of determining the "best" coefficients, $a_k$, is to select those coefficients that minimize the total squared prediction error

$$E = \sum_{n=1}^{N} |e[n]|^2 = \sum_{n=1}^{N} |x[n] - \hat{x}[n]|^2 , \tag{6.28}$$

assuming the sequence $x[n]$ has length $N$.

Several approaches can be used to solve for the $a_k$'s that minimize $E$ in Eq. (??). Perhaps the simplest is to use that MATLAB \ operator for solving simultaneous linear equations. Assuming $N > P$, the linear prediction problem can be posed in matrix form as

$$-\underbrace{\begin{bmatrix} x[1] & \cdots & x[p] \\ x[2] & \cdots & x[p+1] \\ \vdots & \cdots & \vdots \\ x[N-p] & \cdots & x[N-1] \end{bmatrix}}_{\mathsf{X}} \underbrace{\begin{bmatrix} a_1 \\ \vdots \\ a_p \end{bmatrix}}_{\mathsf{a}} + \underbrace{\begin{bmatrix} e[p+1] \\ e[p+2] \\ \vdots \\ e[N] \end{bmatrix}}_{\mathsf{e}} = \underbrace{\begin{bmatrix} x[p+1] \\ x[p+2] \\ \vdots \\ x[N] \end{bmatrix}}_{\mathsf{x}}, \tag{6.29}$$

or compactly as `-Xa+e=x`. This equation can be used to solve for the vector `a` which minimizes the total squared prediction error, `e'*e`. The convention of incorporating the minus sign on the left hand side of Eq. (??) is so that the "prediction-error filter" can be expressed as `e=Xa+x`.

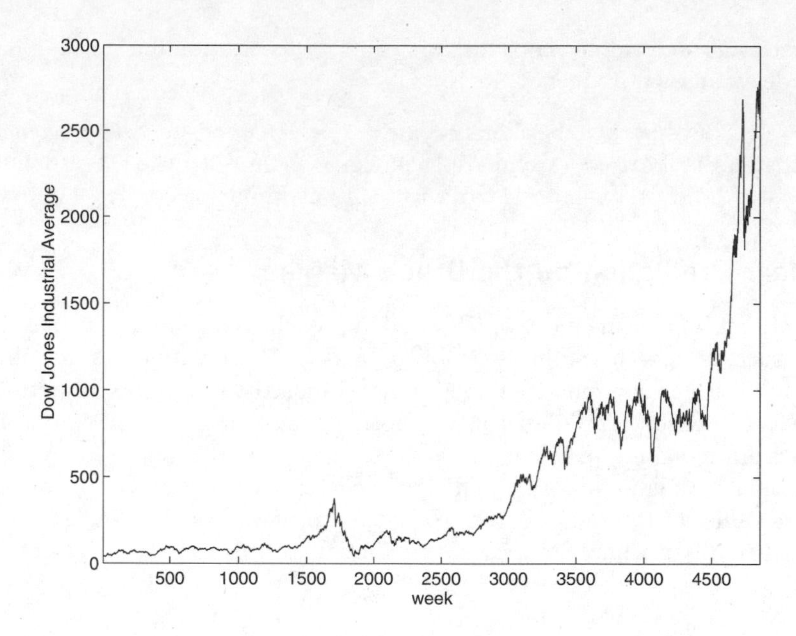


**Figure 6.4.** The Dow Jones Industrial Average over nearly 5000 weeks.

The problems in this exercise will apply linear prediction to the financial data stored in the file `djia.mat`, which is in the Computer Explorations Toolbox. If this file has been loaded correctly, then typing `who` should result in

```
>> who
Your variables are:
djia
```

where `djia` is the Dow Jones Industrial Average (DJIA) index sampled weekly for approximately 94 years. The DJIA for these weeks is plotted in Figure **??**.
In this set of problems, you will attempt to make a fortune by investing with the following strategy:

(i) construct a linear predictor based on past DJIA data;

(ii) use your predictor to guess the value of next week's DJIA based on the past $p$ weeks;

(iii) if the DJIA increases by more than the risk-free interest rate earned by a savings account, you invest all of your money in the DJIA;

(iv) if the DJIA increases by less, you put all of your money in the bank.

You will assume that if you decide put all of your money in the DJIA for the week, then you will earn exactly the gain that was earned in the DJIA. For example, if you had \$1000 in the DJIA at week $n$ and the DJIA at week $n+1$ was given by `djia(n+1)`, then at the end of week $n+1$, you have \$1000`*djia(n+1)/djia(n)`. Also assume that the savings account always earns $r = 3\%$ annual interest, compounded weekly, i.e., your \$1000 would be worth \$1000`*(1+0.03/52)` after one week in the bank.

## Basic Problems

(a). Plot the DJIA data on both a linear and a semi-logarithmic scale. Assuming that you started with \$1000 and invested all of your money in the DJIA, how much money would you have at the end of the investment interval (4861 weeks)? If you had put all of your money in the bank at 3% annual percentage rate (APR), compounded weekly, what rate would you need to achieve the same level of performance? If $r = 0.03$ is the APR, then the bank balance after $N$ weeks from a weekly compounded interest bearing account is equal to $g = (1 + r/52)^N$ times the initial balance.

(b). Assume that $p = 3$ and create the vector `x` and matrix `X` in Eq. (??) from the first decade of data, i.e., use $N = 520$ weeks. The MATLAB \ operator can be used to solve for the vector `a` that minimizes the inner product `e'*e` in Eq. (??). Solve for the linear predictor coefficients using the MATLAB \ operator by `a=-X\x`.

(c). Create the vector of predicted values for the first decade of data using `xhat1=-X*a`. Also create the vector `xhat2` by appropriately using `filter` on the sequence `djia`. Note that the coefficients in the vector `a` are in the reverse of the order required by `filter`. Plot the predicted values on the same set of axes as the actual weekly average. Also determine the total squared error between the predicted and actual values. As a check, do this two ways. First use `e=x+X*a` to compute the prediction error, and then calculate the error by subtracting your predicted sequence `xhat2` from the actual values and make sure that these are the same.

(d). Calculate and plot the total squared prediction error as a function of $p$ for $p = 1, \ldots, 10$. You will have to find the predictor coefficients `a1`, ..., `a10` for each model order $p$, and then calculate each of the prediction errors. What is an appropriate value for $p$, i.e., is there a value of $p$ after which the decrease in prediction error is negligible?

## Intermediate Problems

(e). Given the predictor you designed based on the first decade of data and the model order you have selected from Part ??, you will now test the investment strategy outlined in the introduction. Give yourself \$1000 at the end of the $p$-th week, and make 520 trading decisions based on the output of your predictor. First, determine an upper bound on the amount of money you could make. This would be how much you could make if you were omniscient, i.e., if you knew which direction the stock market was going each week and were always invested in the better of either the bank or the DJIA. Now, as a lower bound, calculate how much money you would make if you left all of your money in the bank and earned a gain of $(1 + 0.03/52)$ each week. As another lower bound, determine how much you would make with the "buy-and-hold" strategy, where you put all your money in the DJIA every week. Finally, calculate how much money you would make with your predictor. What is the equivalent APR that the bank would have had to pay you to achieve the same gain as your predictor?

(f). Now use your prediction strategy on the most recent decade in the data, i.e., the last 520 weeks of the DJIA. Calculate how you did and how each of the bounds perform:

best-possible, all in the bank account, and buy-and-hold. Also calculate the equivalent APR for your predictor.

## Advanced Problems

(g). Compute the maximum gain possible over all of the data. That is, if you knew what the DJIA was going to do each week, and you had the option of making $(1+0.03/52)$ in the bank, or the weekly gain in the DJIA, how much could you make over all 4861 weeks? You may be motivated now to look for additional prediction strategies that could come closer to this maximum gain than the simple linear prediction scheme developed in previous parts. For example, you might try updating your predictor coefficients based on the most recent decade before making each prediction. There are several fast algorithms for doing exactly this, like the recursive least squares (RLS) algorithm[3].

(h). Show that the linear predictor can be used to model the DTFT of the sequence $x[n]$ by analytically demonstrating (using Parseval's relation) that the coefficients $a_k$ are chosen to minimize

$$\frac{1}{2\pi}\int_{-\pi}^{\pi}\left|\frac{X(e^{j\omega})}{\hat{X}(e^{j\omega})}\right|^2 d\omega, \tag{6.30}$$

where

$$\hat{X}(e^{j\omega}) = \frac{1}{1+\sum_{k=1}^{p} a_k e^{-j\omega k}}.$$

Plot the DTFT of the DJIA sequence and the frequency response of the linear predictor on the same set of axes. Since it is not the difference, but rather the ratio, that is minimized, you should see that $\hat{X}(e^{j\omega})$ has the proper shape, but is off by a scale factor $G$. Scale $\hat{X}(e^{j\omega})$ by $G = \sum e^2[n]$ and re-plot the two DTFTs. Can you figure out why this value of $G$ was chosen?

[3]For more on recursive least squares, see *Adaptive Filter Theory*, by S. Haykin.

Chapter 7

# Sampling

The sampling theorem specifies conditions under which a bandlimited continuous-time signal can be completely represented by discrete samples. The resulting discrete-time signal $x[n] = x_c(nT)$ contains all the information in the continuous-time signal so long as the continuous-time signal is sufficiently bandlimited in frequency, i.e., $X_c(j\Omega) = 0$ for $|\Omega| \geq \pi/T$. When this condition is satisfied, the original continuous-time signal can be perfectly reconstructed by interpolating between the samples of $x[n]$. If $x[n]$ satisfies the sampling theorem, it is possible to process $x[n]$ entirely in discrete-time to obtain the samples that would have resulted from sampling $x_c(t)$ at a different rate. This processing is known as sampling rate conversion. Working with sampled data, together with the flexibility of discrete-time systems, leads to a powerful strategy for implementing continuous-time LTI systems known as "discrete-time processing of continuous-time signals." In this technique, a bandlimited continuous-time input signal is sampled, the resulting samples are processed by a discrete-time system, and then the output samples of the discrete-time system are interpolated to give the continuous-time output signal; the overall system is shown in Figure ??.

The exercises in this chapter explore many of the issues involved in sampling and reconstructing signals. Exercise ?? examines the importance of the continuous-time signal being appropriately bandlimited, while Exercise ?? compares two practical approximations to the ideal interpolation filter. Exercise ?? demonstrates how to implement sampling rate conversion using speech sampled at 8192 Hz. In Exercise ??, you will see that for a special class of signals known as bandpass signals, the signal can be sampled at a rate below that given by the sampling theorem and still be perfectly reconstructed. Exercises ?? and ?? explore the half-sample delay and discrete-time differentiator systems, both of which can be analyzed in the context of the discrete-time processing of continuous-time signals.

Note that throughout this chapter $\Omega$ is used to denote the continuous-time frequency vari-

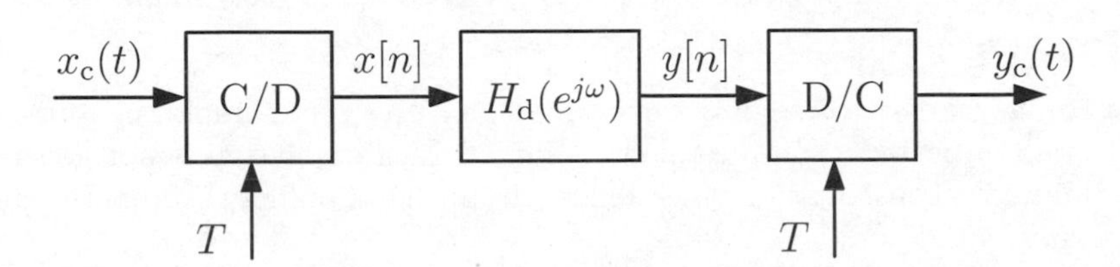


**Figure 7.1.** Discrete-time processing of a continuous-time signal.

able, while $\omega$ is used for the discrete-time frequency variable. This convention should help minimize confusion in the exercises that refer to both time domains.

## 7.1 Aliasing due to Undersampling

This exercise covers the effects of aliasing due to sampling on signals reconstructed by bandlimited interpolation. If a continuous-time signal $x(t)$ is sampled every $T$ seconds, then its samples form the discrete-time sequence $x[n] = x(nT)$. The Nyquist sampling theorem states that if $x(t)$ has bandwidth less than $\Omega_s = 2\pi/T$, i.e., $X(j\Omega) = 0$ for $|\Omega| > \Omega_s/2$, then $x(t)$ can be completely reconstructed from its samples $x(nT)$. The bandlimited interpolation or signal reconstruction is most easily visualized by first multiplying $x(t)$ by an impulse train

$$x_p(t) = \sum_{n=-\infty}^{\infty} x(nT)\,\delta(t - nT)\,.$$

The signal $x(t)$ can be recovered from $x_p(t)$ by filtering $x_p(t)$ with an ideal lowpass filter with cutoff frequency $\Omega_s/2$. Define $x_r(t)$ to be the reconstructed signal given by lowpass filtering $x_p(t)$. If the bandwidth of $x(t)$ is greater than $\Omega_s$, then the samples $x(nT)$ do not completely determine $x(t)$, and $x_r(t)$ will not generally be equal to $x(t)$. In the following problems, you will examine the effects of undersampling a pure sinusoid and a chirp signal.

### Basic Problems

Consider the sinusoidal signal

$$x(t) = \sin(\Omega_0 t)\,.$$

If $x(t)$ is sampled with frequency $\Omega_s = 2\pi/T$ rad/sec, then the discrete-time signal $x[n] = x(nT)$ is equal to

$$x[n] = \sin(\Omega_0 nT)\,.$$

Assume the sampling frequency is fixed at $\Omega_s = 2\pi(8192)$ rad/sec.

(a). Assume $\Omega_0 = 2\pi(1000)$ rad/sec and define `T=1/8192`. Create the vector `n=[0:8191]`, so that `t=n*T` contains the 8192 time samples of the interval $0 \le t < 1$. Create a vector `x` which contains the samples of $x(t)$ at the time samples in `t`.

(b). Display the first fifty samples of $x[n]$ versus $n$ using `stem`. Display the first fifty samples of $x(t)$ versus the sampling times using `plot`. (Use `subplot` to simultaneously display these two plots.)

Note that `plot(t,x)` displays a continuous-time signal given the samples in `x`, using straight lines to interpolate between sample values. While this interpolation is not generally equal to the bandlimited reconstruction which follows from the sampling theorem, it can often be a very good approximation.

To compute samples of the continuous-time Fourier transform of the bandlimited reconstruction $x_r(t)$, use the following function:

```
function [X,f] = ctfts(x,T)
% CTFTS calculates the continuous-time Fourier transform (CTFT) of a
% periodic signal x(t) which is reconstructed from the samples in the
% vector x using ideal bandlimited interpolation.  The vector x
% contains samples of x(t) over an integer number of periods, and T
% contains the sampling period.
%
% The vector X contains the area of the impulses at the frequency
% values stored in the vector f.
%
% This function makes use of the relationship between the CTFT
% of x(t) and the DTFT of its samples x[n], as well as the
% relationship between the DTFT of the samples x[n] and the DTFS of x[n].
N = length(x);
X = fftshift(fft(x,N))*(2*pi/N);
f = linspace(-1,1-1/N,N)/(2*T);
```

This function uses `fft` to calculate the Fourier transform of the reconstructed signal. The M-file `ctfts.m` is provided in the Computer Explorations Toolbox, and should be placed in your MATLABPATH.

(c). Use `[X,f]-ctfts(x,T)` to calculate the continuous-time Fourier transform of the reconstructed signal $x_r(t)$. Plot the magnitude of `X` versus `f`. Is `X` nonzero at the proper frequency values? (Note that almost all the elements in `X` are nonzero, but most are small values due to numerical round-off errors.) Is the phase of `X` correct, assuming that the phase is equal to zero when `X` is nearly zero, i.e., nonzero only due to round-off error?

## Intermediate Problems

You will now consider the effect of aliasing on the reconstructed signal $x_r(t)$.

(d). Repeat Parts ??–?? for the sinusoidal frequencies $\Omega_0 = 2\pi(1500)$ and $2\pi(2000)$ rad/sec. Again, is the magnitude of `X` nonzero for the expected frequencies? Is the phase of `X` correct?

(e). Play each of the sampled signals created in Part ?? using `sound(x,1/T)`. Does the pitch of the tone that you hear increase with increasing frequency $\Omega_0$? Note that, like `plot`, the function `sound` performs interpolation. In essence, your computer converts the discrete-time signal in MATLAB into a continuous-time signal using a digital-to-analog converter, and then plays this continuous-time signal on its speaker.

(f). Now repeat Parts ?? and ?? — do not repeat Part ?? — for the sinusoidal frequencies $\Omega_0 = 2\pi(3500)$, $2\pi(4000)$, $2\pi(4500)$, $2\pi(5000)$, and $2\pi(5500)$ rad/sec. Also play each sample signal using `sound`. Does the pitch of the tone that you hear increase with each increase in the frequency $\Omega_0$? If not, can you explain this behavior?

### Advanced Problems

Now consider the signal

$$x(t) = \sin(\Omega_0 t + \frac{1}{2}\beta t^2), \tag{7.1}$$

which is often called a chirp signal due to the sound it makes when played through a loudspeaker. The "chirp" sound is due to the increasing instantaneous frequency of the signal over time. The instantaneous frequency of a sinusoidal signal is given by the derivative of its phase, i.e., the argument of $\sin(\cdot)$. For the chirp signal, the instantaneous frequency is

$$\begin{aligned}\Omega_{\text{inst}}(t) &= \frac{d}{dt}\left(\Omega_0 t + \frac{1}{2}\beta t^2\right) \\ &= \Omega_0 + \beta t .\end{aligned}$$

Assume for the following problems that $\Omega_s = 2\pi(8192)$ rad/sec:

(g). Set $\Omega_0 = 2\pi(3000)$ rad/sec and $\beta = 2000$ rad/sec$^2$. Store in the vector `x` the samples of the chirp signal on the interval $0 \leq t < 1$.

(h). Use `sound` to play the chirp signal contained in `x`. Can you explain what you just heard?

(i). Determine the approximate time sample at which the chirp signal has its maximum pitch. Given the linear equation for instantaneous frequency and your understanding of aliasing, explain how you could have predicted this time sample.

(j). Store in `x` the samples for the first 10 seconds of the chirp signal. Play the signal using `sound`. Explain how you could have predicted the times at which the played signal has zero (or very low) frequency.

## 7.2 Signal Reconstruction from Samples

This exercise covers the reconstruction of a signal $x(t)$ from its samples $x(nT)$, where $T$ is the sampling period and $n$ is any integer. As discussed in the introduction to Exercise ??, if $x(t)$ has bandwidth less than $\Omega_s = 2\pi/T$, then $x(t)$ can be completely recovered by lowpass filtering the pulse train sampling of $x(t)$:

$$x_p(t) = \sum_{n=-\infty}^{\infty} x(nT)\,\delta(t - nT) .$$

The lowpass filter used to reconstruct $x(t)$ from $x_p(t)$ is

$$H(j\Omega) = \begin{cases} T, & |\Omega| < \Omega_s/2 , \\ 0, & \text{otherwise} , \end{cases}$$

which is an ideal lowpass filter with cutoff frequency $\Omega_s/2$. This filter has the impulse response

$$h_{\text{bl}}(t) = \frac{\sin(\Omega_s t/2)}{\Omega_s t/2}. \tag{7.2}$$

The bandlimited reconstruction is then given by

$$x_{\text{bl}}(t) = \sum_{n=-\infty}^{\infty} x(nT)\, h_{\text{bl}}(t - nT).$$

Whether or not this signal is a "good" reconstruction of $x(t)$ depends on the bandwidth of $x(t)$. As covered in Exercise ??, if $x(t)$ has bandwidth greater than $\Omega_s$, then the reconstruction $x_{\text{bl}}(t)$ generally will not be equal to $x(t)$.
If $x(t)$ has a bandwidth exceeding $\Omega_s$, it still may be possible to recover $x(t)$ from its samples $x(nT)$ if you have additional information about $x(t)$. For instance, if $x(t)$ is known to be piecewise linear, a linear interpolator may be used to reconstruct $x(t)$. The linear interpolation of the samples $x(nT)$ is given by convolving $x_p(t)$ with the following impulse response:

$$h_{\text{lin}}(t) = \begin{cases} 1 - |t|/T, & |t| \leq T, \\ 0, & \text{otherwise}. \end{cases} \tag{7.3}$$

The continuous-time reconstructed signal $x_{\text{lin}}(t) = x_p(t) * h_{\text{lin}}(t)$ is equivalent to connecting the samples $x(nT)$ with straight lines. However, just as bandlimited interpolation may not do a very good job of recovering a signal which is sampled below the Nyquist rate, the linear interpolator may not produce a very good reconstruction if the original signal $x(t)$ is not piecewise linear. As the following problems demonstrate, the performance of any interpolation filter depends upon the characteristics of the original signal $x(t)$.
In the following problems, you will use both bandlimited and linear interpolation to reconstruct the signals

$$x_1(t) = \cos\left(\frac{8\pi t}{5}\right),$$

$$x_2(t) = \begin{cases} 1 - |t|/2, & |t| \leq 2, \\ 0, & \text{otherwise}, \end{cases}$$

from samples obtained at sample times $t = nT$ with $T = 1/2$.

## Basic Problems

(a). Show analytically that both $x_{\text{bl}}(t)$ and $x_{\text{lin}}(t)$ are equal to the sample values $x(nT)$ at the sampling times $t = nT$. Such interpolators are called exact interpolators, since they maintain the exact values of the original signal at the sampling times. Are the bandlimited and linear interpolation filters causal?

(b). Are $x_1(t)$ and $x_2(t)$ bandlimited? If so, with what bandwidth?

(c). Create a vector `ts` which contains the sampling times $t = nT$ on the interval $|t| \leq 4$. Store in the vectors `xs1` and `xs2` the samples of $x_1(t)$ and $x_2(t)$ at the corresponding times in `ts`. Use `stem` to plot `xs1` and `xs2` versus `ts`.

To reconstruct $x_1(t)$ and $x_2(t)$ from these samples, note that the reconstructed signals can only be computed at a finite number of samples in MATLAB. Therefore, you will calculate the interpolated signals only at $t = n/8$ on the interval $|t| \leq 2$. In other words, on the interval $|t| \leq 2$ you will calculate three samples in between every sample contained in `xs1` and `xs2`. The sampling interval of the interpolated signal is thus $T_s = 1/8$.
Another problem encountered in MATLAB is the infinite duration of $h_{\text{bl}}(t)$. One solution is to approximate $h_{\text{bl}}(t)$ by a windowed version of the signal. For the remainder of this exercise, replace $h_{\text{bl}}(t)$ with the finite-length interpolator

$$h_{\text{blf}}(t) = \begin{cases} h_{\text{bl}}(t)\,, & |t| \leq 2\,, \\ 0\,, & \text{otherwise}\,. \end{cases}$$

Call $y_{1\text{bl}}(t)$ and $y_{2\text{bl}}(t)$ the signals given by interpolating the samples of $x_1(t)$ and $x_2(t)$ with the interpolating filter $h_{\text{blf}}(t)$. Similarly, call $y_{1\text{lin}}(t)$ and $y_{2\text{lin}}(t)$ the signals given by interpolating the samples of $x_1(t)$ and $x_2(t)$ with the linear interpolator $h_{\text{lin}}(t)$.

(d). Set `Ti=1/8`, and create a vector of the interpolation times `ti=[-2:Ti:2]`. Store in the vectors `hbl` and `hlin` the values of $h_{\text{blf}}(t)$ and $h_{\text{lin}}(t)$ at the interpolation times. Use `plot` to display these two impulse responses versus `ti`. What are the values of impulse responses at the sample times `ts`? The peak value of each impulse response should be at $t = 0$.

## Intermediate Problems

Note that, because both $h_{\text{lin}}(t)$ and $h_{\text{blf}}(t)$ are nonzero only for $|t| \leq 2$, the interpolated signals on the interval $|t| \leq 2$ are a function only of the samples $x(nT)$ on the interval $|t| \leq 4$. The following problems use `conv` to reconstruct the signals at the interpolation times in `ti` from the samples in `xs1` and `xs2`. However, a lot of bookkeeping is necessary to account for the noncausality of the interpolation filters and to keep track of the samples in `xs1` and `xs2` which correspond to different time samples than those of the interpolation filters `hbl` and `hlin`.

(e). Calculating the interpolated signals at the times in `ti` requires the superposition of a number of shifted versions of `hlin` (or `hbl`), each weighted by the appropriate sample value. This summation is equivalent to a convolution, which can be implemented using the function `conv`. Consider the linear interpolation of the samples of $x_2(t)$. The first step is to make the sample times in `xs2` correspond with those in `hlin`. This can be accomplished by creating the vector

```
>> N = 4*(length(xs2)-1)+1;
>> xe2 = zeros(1,N);
>> xe2(1:4:N) = xs2;
```

The time of each element in `xe2` is given by `te = [-4:Ti:4]`. For each time in `te` equal to a sample time in `ts`, `xe2` contains the corresponding value in `xs2`. Otherwise, `xe2` is zero. Use `stem` to plot `xe2` versus `te`, and compare it with your plot of `xs2`.

(f). Use `conv` to convolve `xe2` with `hlin`. A subset of the output of `conv` contains the linear interpolation at the times `ti`. Remember that the linear interpolator stored in `hlin` corresponds to a noncausal filter, but `conv` assumes that the filter is causal. Refer to Tutorial ?? for an explanation of how `conv` can be used to implement noncausal filters. Extract the desired portion of the output corresponding to $|t| \leq 2$ and store it in the vector `y1lin`. Use `plot` to display `y1lin` versus `ti`. Your interpolation should exactly replicate the values of $x_2(t)$ on the interval $|t| \leq 2$. How could you have predicted this?

(g). Use the same procedure to compute the bandlimited interpolation of the samples of $x_2(t)$. Remember that you only need to compute the interpolation at the times in `ti`. How does this interpolation compare with the linear interpolation? Explain.

(h). Use the same procedure to compute the linear and bandlimited interpolation of the samples of $x_1(t)$. Remember that you only need to compute the interpolation at the times in `ti`. Which interpolation more faithfully reproduces the original signal $x_1(t)$ over the interval $|t| < 2$?

(i). Explain the relative performance of the two interpolators.

## 7.3 Upsampling and Downsampling

In this exercise, you will examine how upsampling and downsampling a discrete-time signal affects its discrete-time Fourier transform (DTFT). If a discrete-time signal was originally obtained by sampling an appropriately bandlimited continuous-time signal, the upsampled or downsampled signal is the set of samples that would have been obtained by sampling the original continuous-time signal at a different sampling rate. For this reason, upsampling and downsampling are often referred to as sampling-rate conversion. Just as with sampling in continuous time, if a discrete-time signal is not sufficiently bandlimited, downsampling may introduce aliasing, which will destroy information. Individually, these operations can only increase or decrease the sampling rate by integer factors, but sampling-rate conversion by any rational factor can be achieved through a combination of upsampling and downsampling.

### Basic Problems

(a). For most of this exercise, you will be working with finite segments of the two signals

$$x_1[n] = \left(\frac{\sin(0.4\pi(n-62))}{0.4\pi(n-62)}\right)^2, \tag{7.4}$$

$$x_2[n] = \left(\frac{\sin(0.2\pi(n-62))}{0.2\pi(n-62)}\right)^2. \tag{7.5}$$

Define x1 and x2 to be these signals for $0 \leq n \leq 124$ using the sinc command. Plot both of these signals using stem. If you defined the signals properly, both plots should show that the signals are symmetric about their largest sample, which has height 1. Analytically confirm your signals have their zero-crossings in the correct locations.

(b). Analytically compute the DTFTs $X_1(e^{j\omega})$ and $X_2(e^{j\omega})$ of $x_1[n]$ and $x_2[n]$ as given in Eqs. (??) and (??), ignoring the effect of truncating the signals. Use fft to compute the samples of the DTFT of the truncated signals in x1 and x2 at $\omega_k = 2\pi k/2048$ for $0 \leq k \leq 2047$ and store the results in X1 and X2. Generate appropriately labeled plots of the magnitudes of X1 and X2. How do these plots compare with your analytical expressions?

(c). Define the expansion of the signal $x[n]$ by $L$ to be the process of inserting $L-1$ zeros between each sample of $x[n]$ to form

$$x_{\mathrm{e}}[n] = \begin{cases} x[n/L], & n = kL, k \text{ integer}, \\ 0, & \text{otherwise}. \end{cases} \tag{7.6}$$

If x is a row vector containing $x[n]$, the following commands implement expanding by L:

```
>> xe = zeros(1,L*length(x));
>> xe(1:L:length(xe)) = x;
```

Based on this template, define xe1 and xe2 to be x1 and x2 expanded by a factor of 3. Also, define Xe1 and Xe2 to be 2048 samples of the DTFT of these expanded signals computed using fft. Generate appropriately labeled plots of the magnitude of these DTFTs. Expanding by $L$ should give a DTFT $X_{\mathrm{e}}(e^{j\omega}) = X(e^{j\omega L})$. Do your plots agree with this?

If you want to increase the sampling rate by $L$, you need to interpolate between the samples of $x_{\mathrm{e}}[n]$ with a lowpass filter with cutoff frequency $\omega_c = \pi/L$. This filter will remove the compressed copies of $X(e^{j\omega})$ located every $2\pi/L$, except the ones centered at $\omega = 2\pi k$. The resulting spectrum is that which would have been obtained if the original bandlimited continuous-time signal had been sampled $L$ times faster. For this reason, the combination of expansion and interpolation is often referred to as upsampling a signal.

## Intermediate Problems

If the row vector x contains the signal $x[n]$, the following MATLAB command will implement downsampling by an integer factor $M$:

```
>> xd = x(1:M:length(x));
```

(d). Based on the DTFTs you found analytically in Part ??, state for both $x_1[n]$ and $x_2[n]$ if the signal can be downsampled by a factor of 2 without introducing aliasing. If downsampling introduces aliasing, indicate which frequencies are corrupted by the aliasing and which are not affected. If the signal can be downsampled without introducing aliasing, sketch the magnitude of $X_\mathrm{d}(e^{j\omega})$, the DTFT of downsampled signal.

(e). Define `xd1` and `xd2` to be the result of downsampling `x1` and `x2` by 2. Define `Xd1` and `Xd2` to be samples of the DTFTs of the downsampled signals computed at 2048 evenly spaced samples between 0 and $2\pi$. Generate appropriately labeled plots of the magnitudes of both DTFTs. Do the plots agree with your sketch(es) from Part ??

(f). A common practice is to process a signal with a lowpass filter whose cutoff frequency is $\pi/M$ before it is downsampled by $M$. This filter is known as an anti-aliasing filter. Even if the signal is not bandlimited to $\pi/M$, the output of the anti-aliasing filter can be downsampled by $M$ without introducing aliasing. If the original signal is not bandlimited to $\pi/M$, the anti-aliasing filter will destroy information, but usually more information is preserved after downsampling than if the signal were downsampled without being processed by the anti-aliasing filter.

In Part ??, you should have found that one of the signals suffered from aliasing when you downsampled it. You can type

```
>> h = 0.5*sinc(0.5*(-32:32)).*(hamming(65)');
```

to define `h` as the impulse response of a lowpass filter with cutoff $\pi/2$. Use the signal which was aliased in Part ?? as the input to this filter, and then downsample the output of the filter by 2. Use `fft` to compute 2048 samples of the DTFT of the downsampled signal. Generate appropriately labeled plots of the magnitude of the DTFT. Based on this plot, determine how much of the original DTFT is preserved.

## Advanced Problems

(g). The method used to downsample `x` in Part ?? keeps the odd-indexed samples because the index of `x` starts at 1. If you change the command to keep the even-indexed samples, the resulting DTFT is significantly different for the signal which suffered from aliasing. Generate a plot of the magnitude of the DTFT that results if you keep the even-indexed samples. Let $X_\mathrm{od}(e^{j\omega})$ be the DTFT of the downsampled sequence that results from keeping the odd-indexed samples and let $X_\mathrm{ev}(e^{j\omega})$ be the DTFT that results from keeping the even-indexed samples. Prove that if the signal downsampled is finite-length, real-valued and symmetric about a sample, at least one of $X_\mathrm{od}(e^{j\omega})$ and $X_\mathrm{ev}(e^{j\omega})$ will be zero at $\omega = \pi$.

(h). For this problem, you will need the two data files `orig.mat` and `orig10k.mat`, which are in the Computer Explorations Toolbox. The file `orig10k.mat` contains a speech signal sampled at 10240 Hz. The effective sample rate of a discrete-time signal can be converted by any rational amount by the correct cascade of upsampling, downsampling, interpolation filters, and anti-aliasing filters. In this problem, you will design a system to process this speech signal to obtain the samples that would have been obtained by sampling at 8192 Hz. If `orig10k.mat` is already in your MATLABPATH, you can load the signal by typing

```
>> load orig10k;
>> who
Your variables are:
x10k
```

Determine the correct integer factors for upsampling and downsampling to convert the sampling rate from 10240 Hz to 8192 Hz. Should the converted signal have more or less samples than `x10k`? Does it matter whether you upsample or downsample first? Are both an interpolating and an anti-aliasing filter required? Implement your system, and play the original and converted signals at 8192 Hz using `sound` to confirm the converted signal. Load `orig.mat`, which contains the original speech signal sampled at 8192 Hz, and confirm that your converted signal sounds the same as the `x` in `orig.mat`.

## ■ 7.4 Bandpass Sampling

This exercise explores some of the issues that arise when sampling a continuous-time signal which has a bandpass Fourier transform. The sampling theorem states that a bandlimited signal must be sampled at a rate that is higher than the bandwidth of the signal for perfect reconstruction to be possible. For real-valued signals which are lowpass in frequency, the sampling frequency is twice the highest frequency in the signal. However, for a large class of bandpass signals, it is often possible to sample the signal at a much slower rate without loss of any information in the signal.
Consider the real-valued narrowband chirp signal, which for $0 \leq t \leq t_0$ is given by

$$x(t) = \cos\left(\Omega_{\min} t + \frac{\Omega_{\max} - \Omega_{\min}}{2t_0} t^2\right). \tag{7.7}$$

The instantaneous frequency of this signal can be found by taking the time-derivative of the phase, i.e., the argument of the cosine, which yields

$$\Omega(t) = \Omega_{\min} + (\Omega_{\max} - \Omega_{\min})t/t_0. \tag{7.8}$$

As you can see, the frequency of this signal increases linearly with time from $\Omega_{\min}$ to $\Omega_{\max}$ in $t_0$ seconds. If the chirp signal $x(t)$ is zero outside $0 \leq t \leq t_0$, then it is approximately bandlimited to $\Omega_{\min} \leq |\Omega| \leq \Omega_{\max}$. For the problems in this exercise, set $\Omega_{\min} = 2\pi(2000)$ rad/sec, $\Omega_{\max} = 2\pi(3000)$ rad/sec, and $t_0 = 0.5$ seconds.

### Basic Problems

Since MATLAB cannot support continuous-time signals, and this exercise will be using both discrete- and continuous-time signals, you will need to maintain a set of samples of the continuous-time axis. It is important to keep track of the difference between the samples used to simulate the continuous-time signal in MATLABand the discrete-time samples that arise from sampling. To simulate the continuous-time axis, create the time vector `t=linspace(0,t0,t0/T)`, for `T=1/8192` and `t0=0.5`.

(a). Create the chirp signal $x(t)$ at the times stored in `t` and store the result in the vector `x`. Using Eq. (??), plot the instantaneous frequency of $x(t)$ over the range $0 \le t \le 0.5$ seconds. Listen to the chirp signal using `sound(x,1/T)`.

(b). Using `fft` and the samples in `x`, compute 8192 samples of the continuous-time Fourier transform (CTFT) of $x(t)$ and store the result in the vector `X`. Plot the magnitude of the CTFT of $x(t)$ versus $-\pi/T \le \Omega < \pi/T$. Be sure to appropriately label the frequency axis.

(c). For the remainder of this exercise, assume that the CTFT of $x(t)$ is bandpass and

$$X(j\Omega) = \begin{cases} 1, & 2\pi(2000) < |\Omega| < 2\pi(3000) \text{ rad/sec}, \\ 0, & \text{otherwise}. \end{cases} \tag{7.9}$$

From the sampling theorem, what is the sampling rate required such that the chirp signal could be sampled without aliasing? Make a sketch of the magnitude of the CTFT that would result from sampling the signal $x(t)$ at a sampling rate of `fs=2000`. Note whether or not any aliasing has occurred, i.e., are any high frequencies mapped to lower frequencies? Also note whether or not the signal $x(t)$ could be completely recovered from this set of samples. If it can be recovered from these samples, indicate how you would do so. If it cannot, indicate why. As a check, create the vector `x1` containing samples of the signal $x_1[n] = x(n/2000)$. Plot the magnitude of the Fourier transform of $x_1(t)$, using

```
>> plot(linspace(-2000,2000,8192),abs(fftshift(fft(x1,8192))));
```

## Intermediate Problems

In these problems, you will use the following strategy to sample the real-valued signal $x(t)$ with `fs=2048`. First, modulate the signal such that the band of frequencies with nonzero energy between 2 kHz and 3 kHz is centered around $\Omega = 0$. Second, lowpass filter the signal, and third, sample the signal at a rate of `fs=2048`.

(d). Use the modulation property of the Fourier transform to determine an analytic expression for the signal $x_m(t)$ which has the same Fourier transform as $x(t)$, but is symmetric about $\Omega = -2\pi(2500)$ rad/sec, i.e., $X_m(j\Omega) = X\big(j(\Omega + 2\pi(2500))\big)$. Create a vector of $x_m(t)$ evaluated at the times stored in `t` and store the result in the vector `xm`. Using `fft`, create an appropriately labeled plot of the magnitude of the CTFT of $x_m(t)$ over the range, $-\pi/T \le \Omega < \pi/T$, for $T = 1/8192$ seconds.

(e). You now need to lowpass filter the signal $x_m(t)$ with an appropriately chosen cutoff frequency that only keeps the band of frequencies centered at zero. Since the signal $x(t)$ is contained in the vector `xm` at an equivalent sampling rate of `fs=8192` Hz, you can perform the lowpass filtering of $x(t)$ with a discrete-time filter. This filter has the appropriate discrete-time cutoff frequency such that when mapped back to continuous time at a sampling rate of `fs`, it corresponds to the desired continuous-time cutoff frequency. Determine this discrete-time cutoff frequency, and store its value in the variable wc. To create a lowpass filter with this discrete-time cutoff frequency, use the impulse response

```
>> h=wc*sinc([-16:16]*wc).*hamming(33)';
```

Create the signal $y_m(t)$ by filtering the signal $x_m(t)$ using ym=filter(h,1,xm). Using fft, create and plot the Fourier transform magnitude of the signal $y_m(t)$ over the range of frequencies, $-\pi/T \leq \Omega < \pi/T$, for $T = 1/8192$ seconds.

(f). Since the signal $y_m(t)$ is bandlimited to $\pm 500$ Hz, it can be sampled with any sampling rate greater than 1 kHz without losing any of the information in the signal. Create samples of the signal $y_m(t)$ at a sampling rate of 2048 Hz from the vector ym. Store the result in the vector ym2. Plot the magnitude of the DTFT of ym2.

## Advanced Problems

To verify that all of the information from the signal $x(t)$ is contained in the samples ym2, you will now reconstruct $x(t)$. This can be done by first using bandlimited interpolation to reconstruct the sequence ym from the samples contained in ym2. You can then modulate the signal to the center frequency of the chirp, therefore reconstructing the signal contained in the vector x.

(g). The first step is to use bandlimited interpolation to increase the sampling rate by a factor of 4. If you have the professional version of MATLAB, the function interp can be used to perform the sample rate conversion from 2048 Hz to 8192 Hz by typing ymhat=interp(ym2,4).

(h). Now recover the signal corresponding to the positive frequency portion of the original bandpass signal $x(t)$ by modulating the signal in the vector ymhat up to the center frequency of the chirp. Use a complex exponential for the modulation and store the result in the vector xm. Use the symmetry properties of the Fourier transform and your knowledge that the signal $x(t)$ is real-valued to recover the negative frequency portion of the signal $x(t)$, and store the sum of the positive and negative frequency components in the vector xm2. Plot the Fourier transform magnitude of xm2 with the frequency axis properly labeled. Verify that you have indeed recovered the signal $x(t)$ at a sampling rate of 8192 Hz by playing xm2 using sound and comparing this to x.

## 7.5 Half-Sample Delay

The effect of multiplying a discrete-time Fourier transform (DTFT) $X(e^{j\omega})$ by an exponential $e^{-j\omega n_d}$ is to shift the signal $x[n]$ in the time-domain. If $n_d$ is a positive integer, this is easily understood as delaying the discrete-time signal to obtain $x[n - n_d]$. When $n_d$ is not an integer, the effect is more difficult to understand, since discrete-time signals are only defined for integer values of their arguments. In this exercise, you will explore the effect of a system whose frequency response is

$$H(e^{j\omega}) = e^{-j\omega/2}, \quad |\omega| < \pi. \tag{7.10}$$

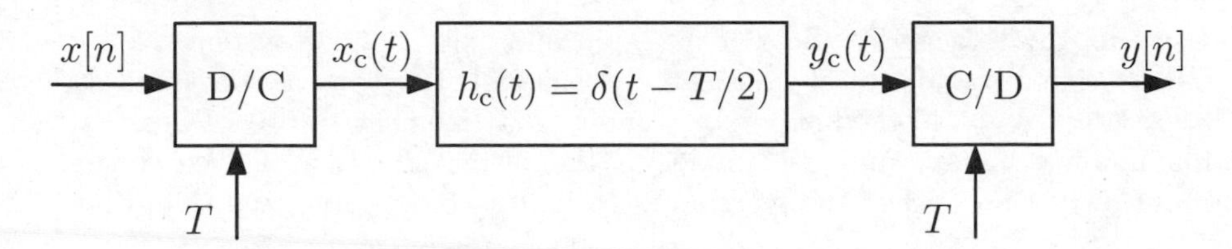


**Figure 7.2.** The half-sample delay of a discrete-time signal $x[n]$ can be constructed by first determining $x_c(t)$, the bandlimited interpolation of $x[n]$. The half-sample delay $y[n]$ is given by sampling $y_c(t)$, a delayed version of the continuous-time signal $x_c(t)$.

The output of this system for the input $x[n]$ is certainly not "$x[n-1/2]$", since discrete-time signals are not defined for noninteger values of their independent variable. In this exercise, you will see that the output of this system, known as a "half-sample delay," can be thought of as samples of a delayed version of the bandlimited interpolation of $x[n]$ as shown in Figure ??.

## Basic Problems

(a). The impulse response of the half-sample delay system in Eq. (??) is

$$h[n] = \frac{\sin(\pi(n - 1/2))}{\pi(n - 1/2)}. \tag{7.11}$$

Define `h` to contain the values of $h[n]$ for the interval $-31 \leq n \leq 32$. Generate a plot of `h` using `stem`.

(b). For the system whose impulse response is stored in `h`, compute the frequency response at 1024 evenly spaced frequencies between 0 and $\pi$ using `freqz`. Generate an appropriately labeled plot of the magnitude of the frequency response. Could you have predicted the value of $H(e^{j\omega})|_{\omega=\pi}$ based on your plot from Part ???

(c). For the half-sample delay system with the frequency response shown in Eq. (??), the group delay for the system is a constant $n_d$. Because the group delay is a constant, all frequencies will be delayed equally as they pass through the system. Use the function `grpdelay` to compute the group delay of the system with the impulse response contained in `h`. Exercise ?? explains the use of the `grpdelay` function. Is $n_d$ an integer for this system? Note that `grpdelay` assumes the system is causal and that `h(1)` $= h[0]$, so that the values returned by `grpdelay` will be larger than expected from Eq. (??).

(d). Use the function `sinc` to define `x` to be the signal

$$x[n] = \frac{\sin(\pi n/8)}{\pi n/8} \tag{7.12}$$

for $-127 \leq n \leq 127$. Compute the output of the LTI system with impulse response `h` to this signal using `filter`, and store the result in `y`.

(e). Use `subplot` to generate plots of the input and output of the system. Note that the output signal has an initial transient. To make a better comparison, discard the first 31 samples of `y`. Note that `x` has a unique maximum at $n = 0$. Does `y` attain its maximum value at a unique sample? Is the output simply a delayed version of the input, or is the effect of the system on the input more complicated than that?

(f). Define `y2` to be the output of the system with impulse response `h` when you use `y` as the input. Effectively, this is cascading two half-sample delay systems together. Use `subplot` to compare `y2` with the original input `x`. Does `y2` have a unique maximum? How are these two signals related? Define the impulse response of the overall system `h2` to be the convolution of `h` with itself using `conv`. Plot `h2` using `stem`. Does this make sense given what you saw in your plot of `y2` and `x`?

### Intermediate Problems

(g). Consider computing $x_c(t)$ as the bandlimited interpolation of $x[n]$ with $T = 1$, i.e., assume the values of $x_c(nT)$ are given by $x[n]$ and between these samples the continuous-time signal is obtained by interpolating with a continuous-time lowpass filter with a cutoff frequency of $\pi$. Write an analytic expression for $x_c(t)$ for the signal $x[n]$ defined in Part ??.

(h). The group delay you found in Part ?? had the units of samples. Converting to continuous time, we have $T = 1$ sec/sample. How many seconds should the continuous-time signal $x_c(t)$ be delayed to introduce a delay equivalent to that generated by the discrete-time system with impulse response given by `h`? Write an analytic expression for $y_c(t)$, the result of delaying the interpolated signal $x_c(t)$ by this amount.

(i). Now that you have defined $y_c(t)$ to be a delayed version of the bandlimited interpolation of the input; you need only resample this signal to compute the half-sample delay. Define `yc` to be the values obtained by sampling $y_c(t)$ every $T = 1$ seconds. Using the `hold` function, generate a plot where the values of `yc` are connected by solid lines using `plot`, while `y` is plotted using `stem`. Other than the initial transient, do the values of `y` fall on the line given by `yc`?

## 7.6 Discrete-Time Differentiation

A differentiator is one example of a continuous-time system that can be implemented using discrete-time processing of continuous-time signals as shown in Figure ??. Using an ideal differentiator can be problematic in many practical situations, as a differentiator will amplify any high-frequency noise in the input signal. For this reason, it is usually desirable to use a bandlimited differentiator, where the cutoff frequency for the bandlimiting is above the highest frequency of the desired input. This exercise considers the design of finite-length impulse response (FIR) filters which can be used in a discrete-time implementation of a bandlimited differentiating filter. The frequency response of a continuous-time differentiating filter is

$$H_c(j\Omega) = j\Omega, \tag{7.13}$$

and that of a bandlimited differentiator with cutoff frequency $\Omega_c$ is

$$H_c(j\Omega) = \begin{cases} j\Omega, & |\Omega| < \Omega_c, \\ 0, & |\Omega| > \Omega_c. \end{cases} \quad (7.14)$$

If the continuous-time signal is appropriately bandlimited, a differentiator can be implemented in discrete-time by 1) sampling the signal, 2) processing the signal in discrete-time and then 3) reconstructing a continuous-time signal from the discrete-time output signal as shown in Figure ??. This process is known as the "discrete-time processing of a continuous-time signal." If this process is implemented using a sampling frequency of $\Omega_s = 2\Omega_c = 2\pi/T$, the corresponding discrete-time transfer function $H_d(e^{j\omega})$ is

$$H_d(e^{j\omega}) = j\left(\frac{\omega}{T}\right), \quad |\omega| < \pi. \quad (7.15)$$

For the remainder of this exercise, you will design FIR approximations to the ideal discrete-time differentiator, assuming that $T = 1$. The impulse response of the ideal discrete-time differentiator is

$$h_d[n] = \begin{cases} 0, & n = 0, \\ \dfrac{(-1)^n}{n}, & n \neq 0. \end{cases} \quad (7.16)$$

Note that $H_d(e^{j\omega})$ is purely imaginary and $h_d[n]$ is antisymmetric about $n = 0$. If $H(e^{j\omega})$ is multiplied by a linear phase term,

$$H(e^{j\omega})e^{-j\alpha\omega}, \quad (7.17)$$

the resulting impulse response will be antisymmetric about $n = \alpha$, assuming $2\alpha$ is an integer. If $\alpha$ is an integer, the impulse response will be $h_d[n-\alpha]$.

## Basic Problems

(a). From Eq. (??) sketch the magnitude and phase of $H_d(e^{j\omega})$ for $-2\pi < \omega < 2\pi$. Remember $T = 1$ throughout this exercise.

(b). Generate and plot $h_d[n]$ for $-256 < n < 256$.

A simple design procedure for a causal FIR differentiator is to first delay $h_d[n]$, then multiply $h_d[n-n_d]$ by a finite-length causal window, $w[n]$. The window should be chosen to maintain the symmetry properties of $h_d[n-n_d]$, so $h[n] = w[n]h_d[n-n_d]$ will retain the linear phase character of Eq. (??). In order to maintain the symmetry of $h[n]$, $w[n]$ must be symmetric about the same point as $h_d[n-n_d]$.

(c). Derive a relation between the delay $n_d$ and the window length $N$ such that $h[n]$ will be causal and have the desired symmetry. Does this relation limit the possible choices of $N$ for a fixed $n_d$?

(d). A rectangular window of length $N$ is defined to be equal to one for $0 \leq n < N$, and equal to zero elsewhere. Construct the impulse response $h[n] = w[n]h_d[n - n_d]$, where $w[n]$ is a rectangular window of length $N = 511$, created by `w=boxcar(511)`. Choose an appropriate value for $n_d$ so that $h[n]$ is causal and store your result in the vector `h`. Use `H=fft(h,1024)` to generate 1024 equally spaced samples of $H(e^{j\omega})$. Plot the magnitude of `H` versus $\omega$ for $-\pi \leq \omega < \pi$. How do you explain the behavior of the magnitude in the vicinity of $\omega = \pm\pi$?

(e). Generate and plot the magnitude of $H(e^{j\omega})$ when $w[n]$ is a rectangular window of length $N = 17$. How does this frequency response compare to that given in Part **??**?

(f). Generate samples on the interval $0 \leq t \leq 100$ of the signal $x_c(t) = \cos(\Omega_c t)$ for $\Omega_c = \pi/10$ and $T = 1$. Store the result in the vector `xd`. Plot $x_d[n] = x_c(nT)$ over the range $0 \leq t \leq 100$.

(g). Take the derivative of the signal $x_c(t)$ by processing the sampled signal $x_d[n]$ using the discrete-time differentiator from Part **??** using `yd=conv(h,xd)`. Since you are implementing the differentiator with a causal filter, and the ideal differentiator is noncausal, your resulting `xd` will be delayed by $n_d$ samples, where $n_d$ is the point of symmetry of the filter, i.e., for $N = 17$, $n_d = 8$. (Refer to Tutorial **??** for a discussion of how to use `conv` to implement noncausal filtering.) Determine analytically the derivative of $x_c(t - n_d)$ and store samples of the derivative in the vector `ytrue`. Plot `xd`, `yd`, and `ytrue` on the same set of axes for $20 \leq t \leq 100$. Does the discrete-time differentiator give you roughly the same answer as the analytically determined derivative at the sampled values?

If you have the function `interp` available from the Signal Processing Toolbox, then `yd2=interp(yd,N)` can be used to compute a finer set of time samples of the signal $y_c(nT/N)$ from `yd`. This will be helpful for comparing the output of the discrete-time differentiator with the analytically determined output. For example, typing

```
>> plot(interp(n*T,N),[interp(xd',N) interp(yd',N) interp(ytrue',N)]);
```

will plot the each of the three signals $x_c(t)$, $y_c(t)$, and $dx_c(t)/dt$ at the set of time samples, $t = nT/N$.

(h). Repeat Parts **??** and **??** for $\Omega_c = \pi/4$, $3\pi/4$, and $5\pi/4$. For which frequencies does the discrete-time differentiator perform the proper differentiation? Could you have predicted this?

## Intermediate Problem

As you have observed, the frequency response $H_d(e^{j\omega})$ in Eq. (**??**) has a discontinuity at $\omega = \pm\pi$. This discontinuity causes large ripples in the frequency response obtained by windowing. The discontinuity can be avoided if $H_d(e^{j\omega})$ and $h_d[n]$ are modified to include a "half-sample delay", i.e.,

$$H_d(e^{j\omega}) = j\omega e^{-j\omega/2}, \qquad -\pi < \omega < \pi, \tag{7.18}$$

$$h_d[n] = \frac{4}{\pi} \frac{(-1)^n}{(2n-1)^2} \,. \tag{7.19}$$

Note that $h_d[n]$ is antisymmetric about "$n = 1/2$" rather than $n = 0$. When $h_d[n]$ is delayed and multiplied by a causal window, you want its symmetry to be maintained. To guarantee this, the window $w[n]$ applied to $h_d[n - n_d]$ must have the symmetry

$$w[n] = w[N - n - 1] \tag{7.20}$$

and the window length $N$ must be even.
For the following problem, consider $h[n] = h_d[n - n_d]w[n]$ with $h_d[n]$ given by Eq. (**??**), and with $w[n]$ having the symmetry in Eq. (**??**). Choose an appropriate value for $n_d$ such that your window is causal.

(i). Generate and plot the magnitude and phase of $H(e^{j\omega})$ for $h[n]$ when $w[n]$ is a rectangular window of length $N = 16$. Explicitly indicate how this compares with what you obtained in Part **??**.

## Advanced Problems

Frequency sampling is an alternative to the window method and was explored in Exercise **??**. Frequency sampling allows the design of filters with even or odd length, using the same ideal frequency response $H_d(e^{j\omega})$. For the next set of problems, use $H_d(e^{j\omega})$ as given in Eq. (**??**). In this approach, you choose $h[n]$ to be the `ifft` of $N$ equally spaced samples of $H_d(e^{j\omega})$ given by

$$\omega_k = \frac{2\pi k}{N}, \qquad 0 \le k \le N - 1 \,.$$

Let `H[k+1]` $= H_d(e^{j\omega_k})$. Consider `h = ifft(H)`. From the definition of the DTFT and its inverse, $h[n]$ will be zero outside the interval $0 \le n \le N - 1$. However, the differentiator impulse response should have odd symmetry about some point (which may not be an integer), so $H_d(e^{j\omega_k})$ may need to be multiplied by an appropriate linear phase factor $e^{-j\alpha\omega_k}$ before the inverse DTFT is computed.
Note: Due to roundoff errors, the output of the function `ifft` may be complex. The impulse response $h[n]$ should be real, so you may need to take the real part of the output of `ifft`. If the `ifft` output contains any significant imaginary part (greater than $10^{-10}$), you have probably made an error.

(j). For $N = 19$, generate and plot $h[n]$ so that

$$H(e^{j\omega_k}) = H_d(e^{j\omega_k})e^{-j\alpha\omega_k} \,, \qquad \omega_k = \frac{2\pi k}{N},$$

for $0 \le k \le N - 1$. Determine an appropriate value of $\alpha$ (if one is necessary) such that $h[n]$ has the proper symmetry. Also from $h[n]$, generate and plot the magnitude and phase of $H(e^{j\omega})$.

(k). Repeat Part **??** for $N = 18$.

(l). You should observe that the frequency response obtained in Part **??** is better, i.e., has less ripple, than in **??**, although the filter had a shorter length. Explain why. Hint: Look at the transition band imposed around $\omega = \pi$ in both cases.

Chapter 8

# Communications Systems

This chapter discusses various techniques used in the transmission of information. Communications systems are used to transmit signals over both short distances, as in local computer networks, and long distances, as in satellite broadcast systems. The Fourier transform, in addition to the time-domain tools that you have learned about, can be used both to design and analyze the performance of these communications systems. Two common methods for processing a signal before transmission are amplitude modulation (AM) and frequency modulation (FM). If $x(t)$ is the signal to be transmitted, which is sometimes called the message signal, then amplitude modulation with a sinusoid is given by

$$y(t) = x(t)\,\cos(\omega_c t)\,. \tag{8.1}$$

The frequency $\omega_c$ is called the carrier frequency of the sinusoid. For discrete-time systems, amplitude modulation with a sinusoidal carrier is given analogously by $y[n] = x[n]\cos(\omega_c n)$. The continuous-time Fourier transform of the modulated signal $y(t)$,

$$Y(j\omega) = \frac{1}{2}\Big(X\left(j(\omega-\omega_c)\right) + X\left(j(\omega+\omega_c)\right)\Big)\,, \tag{8.2}$$

provides some insight into the potential application of sinusoidal amplitude modulation. Assuming that $x(t)$ is bandlimited and the carrier frequency is large enough, sinusoidal amplitude modulation can be used to "move" the message signal $x(t)$ to a frequency band which is more amenable to transmission. For instance, while low-frequency signals near 1 kHz are rapidly attenuated by the atmosphere, frequencies in the range of 1 MHz can travel large distances without attenuation. This is one of the reasons that AM and FM radio signals are transmitted in the range of 500 kHz to 100 MHz, even though the message signals are in the range of 50 Hz to 20 kHz. Modulation can also be used to transmit the messages of multiple users over a single channel. In many systems, the goal is to maximize the number of users who can transmit simultaneously over a channel of finite bandwidth. A modification of sinusoidal amplitude modulation which minimizes the bandwidth occupied by the modulated signal is discussed in Exercise ??. A vector interpretation of sinusoidal amplitude modulation is provided in Exercise ??, while the effects of phase mismatch between the transmitter and receiver are covered in Exercise ?? in the context of speech transmission, e.g., over telephone lines.

Using amplitude modulation with a sinusoidal carrier to transmit simultaneously a number of different message signals in nonoverlapping frequency bands is called frequency-division multiplexing. An alternative method for the transmission of multiple signals is time-division multiplexing. In this case, the message signals occupy nonoverlapping time intervals. As with frequency division multiplexing, the number of users which can be transmitted over a channel is determined primarily by the bandwidth of the message signals and the bandwidth of the transmission channel. The effect of channel bandwidth on time-division multiplexing is covered in Exercise ??. In this exercise, amplitude modulation with a pulse train is used to implement the time-division multiplexing.

In some cases, frequency modulation is preferred to amplitude modulation, as for radio transmitters and receivers. Some basic examples of frequency modulation are covered in Exercise ??.

## 8.1 The Hilbert Transform and Single-Sideband AM

In discrete time, amplitude modulation with a sinusoidal carrier is given by

$$y[n] = x[n]\cos(\omega_c n)\,, \tag{8.3}$$

where the message signal $x[n]$ is assumed to have bandwidth less than $2\omega_c$. The modulated signal has a DTFT with duplicate copies of $X(e^{j\omega})$ centered at $\omega = \pm\omega_c$. This duplication is undesirable if one is trying to maximize the number of users who can transmit simultaneously over a communications channel. A naive solution is to replace the sinusoidal carrier with the complex exponential carrier $e^{j\omega_c n}$. However, the resulting modulated signal $x[n]e^{j\omega_c n}$ has an imaginary component, and cannot be transmitted over a real channel. Single-sideband (SSB) modulation is a feasible solution which is equivalent to filtering $y[n]$ prior to transmission with an ideal lowpass filter of cutoff frequency $\omega_c$. The filtered signal occupies the same frequency bandwidth as does $x[n]$, and $x[n]$ can be completely recovered from the transmitted signal.

In this exercise, you will consider an alternative method for constructing SSB signals. This method makes use of the Hilbert transform. The frequency response of an ideal Hilbert transform is given by

$$H(e^{j\omega}) = \begin{cases} -j\,, & 0 \le \omega < \pi\,, \\ j\,, & -\pi \le \omega < 0\,. \end{cases} \tag{8.4}$$

Due to the phase of $H(e^{j\omega})$ in Eq. (??), the Hilbert transform is also called a 90°-phase shifter, and it has many practical applications beyond single-sideband modulation[1].

### Basic Problems

The impulse response of the ideal Hilbert transform is both noncausal and has infinite length. In these problems you will consider a shifted and windowed version of the impulse response of the ideal Hilbert transform. The windowing provides an impulse response with finite length, and the shifting is required to make a causal approximation of the impulse response.

[1]For more on Hilbert transforms, see *Signals and Systems: Continuous and Discrete* by Ziemer, Tranter, and Fannin or the more advanced text *Discrete-time Signal Processing* by Oppenheim and Schafer with Buck.

(a). Based upon Eq. (??), what symmetry does the impulse response $h[n]$ of the ideal Hilbert transform possess?

(b). Consider the system with frequency response $H_\alpha(e^{j\omega}) = e^{-j\alpha\omega}H(e^{j\omega})$, where $H(e^{j\omega})$ is defined in Eq. (??). Sketch the magnitude and phase of this system.

(c). Derive the impulse response $h_\alpha[n]$ of the system with frequency response $H_\alpha(e^{j\omega})$. If $\alpha$ is an integer, about which value of $n$ is $h_\alpha[n]$ symmetric?

(d). For $\alpha = 20$, store the values of $h_\alpha[n]$ on the interval $0 \leq n \leq 40$ in the vector `h`. Note that $\alpha$ is chosen such that `h(n+1)` is symmetric about the index `n=20`. Use `stem` to plot the samples in `h` versus $n$.

(e). Use `fft(h,256)` to obtain samples of the DTFT of the windowed and shifted impulse response. Plot the magnitude and phase of this Fourier transform, and be sure to appropriately label the frequency axes. When plotting the phase, make sure to first use `fftshift` to reorder the frequency response samples, and then use `unwrap` to obtain the continuous phase response. How do the magnitude and phase compare with that of the ideal Hilbert transform?

Now you will use the shifted and windowed Hilbert transform to filter the sinusoidal signal $x[n] = \sin(\pi n/8)$.

(f). What is the response of the ideal Hilbert transform to the input $\sin(\pi n/8)$? If $\alpha$ is an integer, what is the response of the system $H_\alpha(e^{j\omega})$ to the input $\sin(\pi n/8)$?

(g). Store in `x` the values of $x[n]$ on the interval $0 \leq n \leq 128$. Use `conv(h,x)` to approximately implement the convolution $h_\alpha[n] * x[n]$ on the interval $20 \leq n \leq 148$. (The approximation is due to truncation of both the input signal $x[n]$ and the impulse response $h_\alpha[n]$.) Store these samples in the vector `xh`. Plot `x` and the samples of the shifted Hilbert transform $h_\alpha[n] * x[n]$ on the same set of axes. Ignoring the transients which occur due to the windowing of the input signal, how does `xh` compare to the response predicted in Part ??? Explain any discrepancies.

(h). The convolution $h_\alpha[n] * x[n]$ is a shifted version of the Hilbert transform of $x[n]$, and you must account for this shift. The vector `xh` contains both samples of $h_\alpha[n] * x[n]$ on the interval $20 \leq n \leq 148$ and samples of $h[n] * x[n]$ on the interval $0 \leq n \leq 128$. Plot `x` and the samples of the Hilbert transform $h[n] * x[n]$ stored in `xh` on the same set of axes. Ignoring the transients in the computed Hilbert transform, use the signal derived in Part ?? to verify that you have properly implemented the Hilbert transform.

Now that you can implement a Hilbert transform, you can use it to construct a single-sideband modulation system.

## Intermediate Problems

A system for computing the SSB modulation of a signal $x[n]$ is shown in Figure ??. The signal $y[n]$ contains the single-sideband modulation of the signal $x[n]$. You will implement

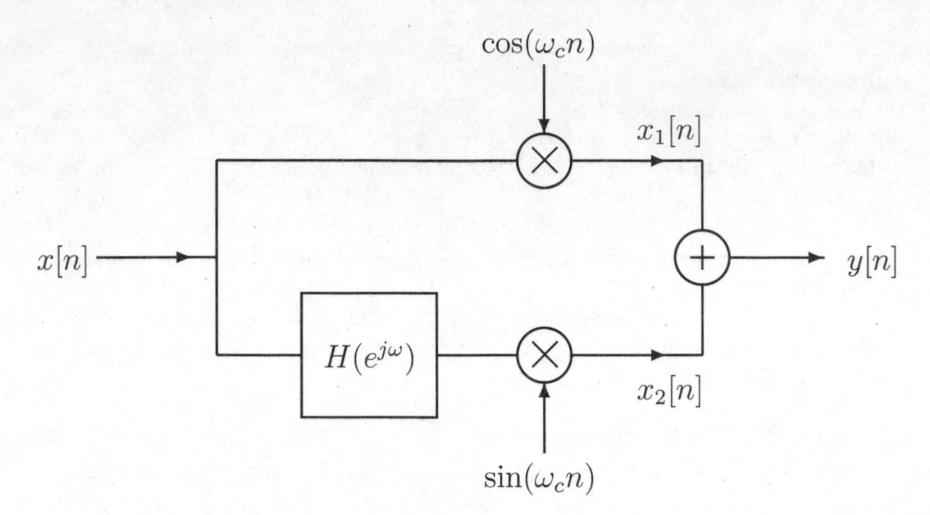


**Figure 8.1.** A system which uses the Hilbert transform $H(e^{j\omega})$ and sinusoidal modulation to obtain the single-sideband modulation of $x[n]$.

this system for the input signal

$$x[n] = \begin{cases} \dfrac{\sin(\pi(n-32)/4)}{\pi(n-32)/4}, & 0 \le n \le 64, \\ 0, & \text{otherwise}. \end{cases} \tag{8.5}$$

Assume that the carrier frequency is $\omega_c = \pi/2$.

(i). Store the nonzero elements of $x[n]$ in the vector `x`. Use `fft(x,256)` to obtain 256 samples of the DTFT of $x[n]$. Plot the magnitude of the DTFT and be sure to appropriately label the frequency axis. Determine the approximate frequency bandwidth occupied by the signal $x[n]$.

(j). Store in `x1` the nonzero samples of $x[n]\cos(\omega_c n)$. Use `fft` to compute 256 samples of the DTFT of $x_1[n] = x[n]\cos(\omega_c n)$. Plot the magnitude of the DTFT and describe its relation to the DTFT of $x[n]$ plotted in Part ??.

(k). Use `conv(h,x)` to compute the Hilbert transform of the signal $x[n]$ on the interval $0 \le n \le 64$. Store this signal in the vector `xh`. Note that selecting the vector `xh` from the output of `conv(h,x)` requires some tedious but necessary bookkeeping on your part. Remember that the filter stored in `h` is a delayed version of the Hilbert transform, and that `conv` computes values of the convolution $h_\alpha[n] * x[n]$ outside the interval $0 \le n \le 64$ by assuming that $x[n]$ is zero outside the interval $0 \le n \le 64$. Hint: The point of symmetry in the vector `xh` should be the same as the point of symmetry in the vector `x`.

(l). Use `fft(xh,256)` to compute samples of the DTFT of the signal stored in `xh`. Plot the magnitude of this DTFT.

(m). Store in `x2` the values of $x_2[n] = x_{\mathrm{h}}[n]\sin(\omega_c n)$ on the interval $0 \leq n \leq 64$, where `xh` contains $x_{\mathrm{h}}[n]$ on this interval. Use `fft` to compute 256 samples of the DTFT of the signal stored in `x2`. Plot the magnitude of the DTFT of `x2` and describe its relation to the DTFT magnitude of $x_{\mathrm{h}}[n]$ plotted in Part ??.

(n). You have now synthesized the two signals necessary to form the modulated signal $y[n]$, which is the output of the system in Figure ??. Store in `y` the values of this signal on the interval $0 \leq n \leq 64$. Use `fft(y,256)` to compute samples of the DTFT of the signal stored in `y`. Plot the magnitude of this DTFT and appropriately label the frequency axis. This plot should illustrate that `y` contains the SSB modulation of $x[n]$ in Eq. (??). Over what frequency range is the DTFT essentially zero? Does the signal stored in `y` require approximately the same bandwidth as $x[n]$? (In this case, define bandwidth as the sum of the widths of the frequency intervals occupied by the signal.)

### Advanced Problem

With a little more effort, you can demodulate the signal $y[n]$ to recover $x[n]$. The first step is to modulate the single-sideband signal stored in $y[n]$ with a sinusoid. What is the correct frequency of this sinusoid? Plot the DTFT of this signal to make sure that $x[n]$ can be safely recovered.

If $y_{\mathrm{m}}[n]$ is the result of modulating $y[n]$ with a sinusoid, then the second step is to lowpass filter the modulated signal. For the lowpass filter, you can use the windowed sinc function

$$h_{\mathrm{lp}}[n] = \begin{cases} A\,\dfrac{\sin(\omega_0 n)}{\omega_0 n}, & -32 \leq n \leq 32\,, \\ 0, & \text{otherwise}\,, \end{cases} \tag{8.6}$$

where $\omega_0$ is the cutoff frequency of the lowpass filter and $A$ is the appropriate normalization. What is the approximate range of cutoff frequencies that you can use to recover $x[n]$? After selecting an appropriate cutoff frequency and normalization, use this lowpass filter to recover $x[n]$ on the interval $0 \leq n \leq 64$. Compare the recovered signal with the original signal stored in `x`. The difference should be very small—on the order of $10^{-3}$.

## ■ 8.2 Vector Analysis of Amplitude Modulation with Carrier

A drawback of the sinusoidal amplitude modulation given by Eq. (??) is that the carrier sinusoid at the receiver must be in phase with the carrier sinusoid at the transmitter. For asynchronous transmission, the receiver does not know and instead must estimate the phase of the transmitter. A method for asynchronous AM transmission is to send the signal

$$y(t) = (A + x(t))\cos(\omega_c t),$$

which is known as sinusoidal AM with carrier, since the modulated signal $y(t)$ contains the carrier $A\cos(\omega_c t)$. If $A + x(t) > 0$, demodulation of the message signal can be performed by tracking the envelope of the transmitted signal $y(t)$. This exercise examines sinusoidal AM with carrier from a vector perspective.

## Basic Problems

This set of problems considers the simple case of asynchronous AM modulation, where the message signal $x(t)$ is a sinusoid of frequency $\omega_m$ and amplitude $m$, i.e.,

$$y(t) = \big(A + m\cos(\omega_m t)\big)\cos(\omega_c t).$$

(a). Sketch the continuous-time Fourier transform (CTFT) of the signal $y(t)$. Assume that $\omega_c \gg \omega_m$ and that $A > m$.

(b). For $A = 1$, $m = 0.5$, $\omega_c = 40\pi$, and $\omega_m = 10\pi$, store in the vector `y` the values of $y(t)$ at the time samples in `t=linspace(0,1,500)`. Plot `y` versus `t`. Store in `y2` the signal $y(t)$ that results from setting $m = 1$ and plot `y2` versus `t`. For what values of `m` can an envelope detector be used to demodulate the message signal, $x(t) = m\cos(\omega_m t)$?

The signal $y(t)$ can be written as a sum of three terms:

$$y(t) = A\cos(\omega_c t) + \frac{Am}{2}\Big(\cos\big((\omega_c + \omega_m)t\big) + \cos\big((\omega_c - \omega_m)t\big)\Big).$$

By noting that $\cos(\theta)$ is the real part of $e^{j\theta}$, each of the cosines in $y(t)$ can be viewed as the projection of the vectors $Ae^{j\omega_c t}$, $(Am/2)e^{j(\omega_c+\omega_m)t}$, and $(Am/2)e^{j(\omega_c-\omega_m)t}$ onto the real axis. The first term can be thought of as a vector of length $A$ rotating about the origin at a rate of $\omega_c$ rad/sec. The next two terms can then be viewed as rotating in opposite directions about the tip of this vector at a rate of $\omega_m$ rad/sec. The overall rotation rates about the origin of these smaller vectors are $(\omega_c+\omega_m)$ and $(\omega_c-\omega_m)$ rad/sec, respectively. The signal $y(t)$ is then the projection of the vector sum onto the real axis.

(c). Change $m$ to 0.5. For the time samples in `t`, store the three vectors $Ae^{j\omega_c t}$, $(Am/2)e^{j(\omega_c+\omega_m)t}$, and $(Am/2)e^{j(\omega_c-\omega_m)t}$ in the MATLAB vectors `y1`, `y2`, and `y3`, respectively. If you have constructed these vectors properly, then executing the code

```
>> plot([0 real(y1(1))],[0 imag(y1(1))],'k-');
>> hold on;
>> plot(real([y1(1) y1(1)+y2(1)]),imag([y1(1) y1(1)+y2(1)]),'r-');
>> plot(real([y1(1) y1(1)+y3(1)]),imag([y1(1) y1(1)+y3(1)]),'g-');
```

will plot the first vector starting at the origin and the two smaller vectors starting at the tip of this vector.

## Intermediate Problems

(d). Fill in the missing parts of the following code, which displays the first 50 consecutive vectors representing $y(t)$. The code fragment `<<<A>>>` specifies the location of the tip of the small vector rotating at $\omega_m + \omega_c$ rad/sec, and `<<<B>>>` specifies the location of the tip of the three vector sum.

```
>> for k=1:50
     plot(real([<<<A>>>]),imag([<<<A>>>]),'ro');
```

```
    hold on
    plot(real([y1(k)+y3(k)]),imag([y1(k)+y3(k)]),'go');
    plot(real([y1(k) <<<A>>>]),imag([y1(k) <<<A>>>]),'r--');
    plot(real([y1(k) y1(k)+y3(k)]),imag([y1(k) y1(k)+y3(k)]),'g--');
    plot(real([0 <<<B>>>]),imag([0 <<<B>>>]),'k-');
    plot(real([<<<B>>>]),imag([<<<B>>>]),'kx');
    axis([-1.5 1.5 -1.5 1.5]);
    axis('square')
    hold off
    pause(.1)
end
```

Use your completed code to make plots for `k=[5 10 15 20]`.

(e). Add a line of code to `for` loop given in Part ?? that stores in the vector `projy` the projection of the vector sum onto the real axis. Also add a line of code to the loop that plots an `'x'` on the real axis at this location. Plot `projy` versus `t` and compare this to your plot from Part ??.

## 8.3 Amplitude Demodulation and Receiver Synchronization

This project explores some of the issues involved in the sinusoidal amplitude modulation and demodulation of speech signals, including the synchronization between the transmitter and receiver. The general form for a sinusoidal AM signal is given by

$$y(t) = x(t)\cos(\omega_c t),$$

where $x(t)$ is called the message signal and $\omega_c$ is the carrier frequency of the AM signal. At the receiver, the message signal $x(t)$ can be recovered through a technique called synchronous AM demodulation via

$$\begin{aligned} w(t) &= 2y(t)\cos(\omega_c t)\,, \\ &= 2x(t)\cos^2(\omega_c t)\,, \\ &= x(t)\big(1+\cos(2\omega_c t)\big)\,. \end{aligned}$$

To recover $x(t)$, the signal $w(t)$ can be lowpass filtered to eliminate the component of the spectrum centered about $2\omega_c$. A potential difficulty in AM systems of this form is that the receiver must have a local oscillator that is synchronized with the transmitter. If there is a phase difference between the transmitter and receiver oscillators, then some demodulation degradation can occur. For example, if the carrier signal is $\cos(\omega_c t)$ and the demodulator signal is $\cos(\omega_c t+\phi)$, then

$$\begin{aligned} w(t) &= x(t)\cos(\omega_c t)\cos(\omega_c t+\phi)\,, \\ &= x(t)\big(\cos(\phi)+\cos(2\omega_c t+\phi)\big). \end{aligned}$$

Depending upon the phase mismatch between the receiver and transmitter, $x(t)\cos(\phi)$ can in fact be zero for all $t$. This issue is explored in more detail in the problems that follow.

The synchronization problem can be avoided if the message signal $x(t)$ is always positive, i.e., $x(t) > 0$. In this case, an envelope detector can be used to demodulate the signal $x(t)$ without any phase synchronization. In such asynchronous AM systems, a positive constant $A$ is often added to the message signal to ensure its positivity, i.e.,

$$y(t) = (A + x(t))\cos(\omega_c t)\,, \tag{8.7}$$

where $A$ is chosen such that $A + x(t) > 0$.

## Basic Problems

These problems cover some of the issues that arise when using synchronous AM modulation to transmit a speech signal. The speech signal you will use is contained in the file `origbl.mat`, which is in the Computer Explorations Toolbox. This file can be loaded by typing `load origbl`. Since the speech is sampled at `fs=8192` Hz, you will use a set of time samples at this rate to simulate continuous-time signals, i.e., `t=[0:1/fs:(N-1)/fs]`, where `N=length(x)`. Verify that you have loaded the speech signal correctly by executing the command `sound(x,8192)`. You should hear a slightly muffled phrase "line up". The speech may sound muffled to you since it has been bandlimited to 1000 Hz.

(a). Use `fft` to approximately compute 8192 samples of the CTFT of `x` (see Exercise 4.2) and store your results in `X`. Make an appropriately labeled plot of the magnitude of the CTFT of `x` to verify that it is indeed bandlimited to about 1000 Hz.

(b). Assuming that $A = 0$ in Eq. (??), amplitude modulate the speech signal in `x` at a carrier frequency of `fc=1500` Hz and store the modulated speech in the vector `y`. Make an appropriately labeled plot of the magnitude of the CTFT of `y`.

(c). Assuming that the demodulator has a perfectly synchronized carrier signal at `1500` Hz, create the demodulated signal $w(t) = y(t)\cos(\omega_c t)$ and store the result in the vector `w`. Use `fft` to compute 8192 samples of the CTFT of `w` and make an appropriately labeled plot of its CTFT magnitude.

(d). Analytically determine the impulse response $h(t)$ of an ideal lowpass filter with cut-off frequency `1500` Hz. Store in a row vector `h` the first 41 samples of $h(t - T)$, for `T=40/fs/2`. Multiply your samples by a hamming window of length 41 by executing `h=h.*hamming(41)'`. Plot the impulse response `h` versus `t(1:41)`. Make an appropriately labeled plot of the CTFT magnitude of `h` by executing

```
>> H=fftshift(fft(h,8192));
>> df = fs/8192;
>> plot([-fs/2:df:fs/2-df],abs(H))
```

## Intermediate Problems

(e). Filter the signal `w` with the impulse response stored in `h` using `filter` and store the result in the vector `z`. Use `fft` to compute 8192 samples of the CTFT and make an appropriately labeled plot of the CTFT magnitude of `z`. Verify that you have properly demodulated the speech signal by playing it using `sound`.

(f). If the receiver did not have a perfectly synchronized oscillator, the demodulated speech signal may be degraded. Demodulate the signal using `w=y.*cos(2*pi*fc*t+phi)` for `phi` equal to 0, $\pi/4$, and $\pi/2$. For each of these possible phase errors, make an appropriately labeled plot of the CTFT magnitude of `w` using `fft`. How will these phase synchronization errors affect the AM demodulation performance? For what value of `phi` will the AM demodulation as described above no longer recover the speech signal or a scaled version of it?

## Advanced Problems

The remaining problems will assume that there is a positive constant $A = 10$ added to the signal $x(t)$ prior to modulation. This extra sinusoidal component in the received signal can then be used to recover the phase of the modulating signal.

(g). For this problem, assume that the transmitter and receiver are out of phase by `phi=pi/4` and create the modulated signal `y=(A+x)'.*cos(2*pi*fc*t+phi)`. A phase-locked-loop (PLL) is often used in practice for "locking" the phase of the receiver to that of the transmitter. Although PLL's are often implemented in hardware, the code fragment shown below can be used to estimate the phase of the received signal `y`. The vector `phihat` contains an estimate of the phase of the carrier and `chat(k)` is an estimate of the carrier at time step `t(k)`.

```
for k=1:N-1,
   chat(k)=A*cos(2*pi*fc*t(k)+phihat(k));
   phihat(k+1)=phihat(k)+alpha*(y(k)-chat(k));
end
```

Find a value of `alpha` and initial value for `phihat(1)` so that this discrete-time PLL "locks" into the phase of the transmitter. Plot the phase difference between the transmitter and receiver versus time. Does `phihat` have to be exactly equal to `phi` for the PLL to be in "lock"? Use the estimated phase of the carrier to demodulate the speech signal and verify your results by playing the speech using `sound`.

(h). In addition to an initial phase uncertainty between the transmitter and receiver, there is often a random phase drift as well. Such phase discrepancies have a variety of causes including component variations between the two systems or changes in the propagation path from source to receiver. For this part, assume that the phase of the carrier as viewed by the receiver is a sum of random phase jitter and is given by `phin=cumsum(randn(1,N)*0.01)`, and construct the received signal

```
>> y=(A+x)'.*cos(2*pi*fc*t+phin);
```

Plot `phin` versus `t` to see the phase drift between the transmitter and receiver. Use a PLL as described in Part ?? to recover the speech signal and verify your results using `sound`. The ability of the PLL to lock onto the phase of the received signal is a function of the RMS value of the phase jitter (which in this example is set to 0.01), the adaptation parameter `alpha`, and the amplitude $A$ of the additional carrier. Vary each of these parameters to explore how they affect the ability of the PLL to lock onto the phase of the transmitted carrier.

## 8.4 Intersymbol Interference in PAM Systems

In a pulse amplitude modulation (PAM) system, a discrete-time message is sent over a communication channel by modulating a periodically repeated pulse shape. Depending on the bandwidth limitations of the channel and the time characteristics of the pulse shape, each of the message samples transmitted by modulating the pulse shape may be received without interference between the adjacent message samples. However, if the bandwidth limitations of the transmission channel result in temporal spreading of the transmitted pulses, then significant intersymbol interference (ISI) can occur. This exercise explores some of the issues within a PAM system that give rise to ISI.

For this exercise, consider a simple binary communication system in which a "one" is represented by transmitting the pulse $p(t)$ and a "zero" is represented by transmitting the negative of the pulse, i.e., $-p(t)$. This is called antipodal signaling. The pulse $p(t)$ that you will use throughout this exercise is given by

$$p(t) = \begin{cases} 1 - |t|/T\,, & |t| < T\,, \\ 0\,, & \text{otherwise}\,. \end{cases} \tag{8.8}$$

The pulses are transmitted at a rate of $f_b$ pulses per second, referred to as the bit rate of the binary PAM system. For the remainder of this exercise, assume that $T = 0.1$. You will consider two possible values for the bit rate, `fb=1/T` and `fb=1/(2T)`.

### Basic Problems

(a). To simulate the continuous-time signals, you will need a set of time samples over which the signals can be plotted. Generate the pulse $p(t)$ at the time samples `[-T:dt:T]` for `dt=T/10` and store the result in the vector `p`. Plot `p` over this range.

(b). Determine the maximum value of $f_b$ so that there is no intersymbol interference in the transmitted waveform. That is, what is the maximum value of $f_b$ such that N pulses in the signal $y(t) = \sum_{k=1}^{N} p(t - k/f_b)$ do not overlap? What is the maximum value of $f_b$ such that samples taken at the bit rate and located at the pulse peaks are not affected by intersymbol interference, i.e., $y(n/f_b) = \sum_{k=1}^{N} p\big((n-k)/f_b\big) = p(0)$?

An eye diagram is a commonly used tool for visually evaluating pulse shapes and determining the degree of intersymbol interference that will result. An eye diagram is displayed by generating a random binary sequence at the bit rate and displaying the resulting signal on an oscilloscope with the bit rate equal to an integer multiple of the oscilloscope's horizontal sweep rate. All of the pulse shapes simply overlap one another on the scope. An eye diagram is a useful tool for determining the ability of a receiver to properly decode the bit stream from the received waveform. In the eye diagram, you will see each of the possible bit levels and the transitions between output levels. A receiver might simply sample the received signal at the bit rate and have to decide whether the signal corresponds to a zero or a one based on this value. The relative separation of these output levels when viewed at times corresponding to the bit rate will indicate how sensitive a receiver will be to noise that may be added to the signal. If the separation is large, then a large amount of noise would be required to induce a bit error. The sensitivity of the signal to timing errors is also evident from the eye diagram. Since the pulses tend to have an open area separating the signal levels sampled at the bit rate, the eye diagram often looks like an eye. If the signal is sensitive to additive noise, the eye is closed from the top and bottom. If the signal is sensitive to timing errors, then the eye will start to close from the sides. The function `eyediagram` given below plots an eye diagram for a signal `y` at a sweep rate of `fb` and sampling period `dt`. The M-file `eyediagram.m` is contained in the Computer Explorations Toolbox, which is available from The MathWorks at the address listed in the Preface.

```
function eyediagram(y,fb,dt)
% eyediagram(y,fb,dt) Plots the eye diagram associated with the PAM
% system using the signal y at a sweep rate fb and at sampling period
% of dt.

tindcs=[0:dt:2/fb];
hold off
for k=4:length(y)*dt*fb-1-4,
  indcs=[ceil((k-1)/fb/dt):ceil((k+1)/fb/dt)]+1;
  plot(tindcs,y(indcs));
  hold on
end
hold off
```

For the following problems, assume `N=20`:

(c). For the pulse `p`, sketch the eye diagram for each of the following cases:

(i) $f_b = 1/T$ and a sequence of all "one"s;

(ii) $f_b = 1/T$ and an alternating sequence of "one"s and "zero"s;

(iii) the two cases above with $f_b = 1/(2T)$;

(iv) a random sequence of "one"s and "zero"s at $f_b = 1/T$;

(v) a random sequence of "one"s and "zero"s at $f_b = 1/(2T)$.

(d). Create the signal $y(t)$ that corresponds to `N` "one"s at the bit rate of $f_b = 1/T$. This can be accomplished by creating an impulse train at the bit rate, and convolving the result with the pulse shape. For example, the code fragment

```
>> y=zeros(1,N*T/dt);
>> y(1:(1/fb)/dt:N*T/dt)=ones(1,N);
>> y=conv(y,p);
>> t=[0:dt:(length(y)-1)*dt];
```

places `N` copies of the pulse stored in `p` at the bit rate `fb`. Create the vectors `y` and `t` as specified above, and plot `y` versus `t` using `stem`. Use the function `eyediagram` to plot the eye diagram corresponding to your sketch from Part ??-i. How do they compare?

(e). Create the vector `y2` containing the `N` pulses corresponding to an alternating sequence of "one"s and "zero"s at a bit rate of $f_b = 1/T$. Plot `y2` versus `t` using `stem`. Use `eyediagram` to plot the eye diagram for `y2`. How does this compare with your sketch from Part ??-ii?

(f). Create the vector `y3` containing the pulses corresponding to a sequence of `N/2` "one"s at a bit rate of $f_b = 1/(2T)$. Plot `y3` versus `t` using `stem`. Use `eyediagram` to plot the eye diagram for `y3`. How does this compare with your sketch from Part ??-iii?

(g). Create the vector `y4` containing the `N/2` pulses corresponding to an alternating sequence of "one"s and "zero"s at a bit rate of $f_b = 1/(2T)$. Plot `y4` versus `t` using `stem`. Use `eyediagram` to plot the eye diagram for `y4`. How does this compare with your sketch from Part ??-iii?

## Intermediate Problems

(h). Use `rand` to create a vector of `N=100` random bits, i.e., create the vector `a` containing 100 entries which are randomly chosen to be either "one" or "zero". This vector can be created by

```
>> a = rand(1,N) > 0.5;
```

Create the vector `y5` containing the `N=100` pulses corresponding to the sequence of "ones" and "zeros" in `a` at a bit rate of $f_b = 1/T$. Plot `y5` versus time using `stem`. Use `eyediagram` to plot the eye diagram for `y5`. How does this compare with your sketch from Part ??-iv?

(i). Create the vector `y6` containing the `N=100` pulses corresponding to the sequence of "one"s and "zero"s in `a` at a bit rate of $f_b = 1/(2T)$. Plot `y6` versus time using `stem`. Use `eyediagram` to plot the eye diagram for `y6`. How does this compare with your sketch from Part ??-v?

## Advanced Problems

In addition to the pulse shape and relative spacing in the transmitted signal, the bandwidth limitations of the transmission channel can also lead to ISI in PAM systems. Consider transmitting the signal $y(t)$ as defined previously over a channel with impulse response $h(t)$. At the receiver, any temporal spreading of the pulse shape $p(t)$ may also impact the eye diagram for the PAM system. For the remaining problems in this exercise, consider the channel to be a bandlimited linear phase channel with impulse response defined by one of

```
>> h1=hamming(41)'.*sinc((t(1:41)-2*T)*2/T);
```

```
>> h2=hamming(41)'.*sinc((t(1:41)-2*T)/T);
```

(j). Create the signal w5 corresponding to the signal that the receiver would observe if the signal y5 were passed through a channel with impulse response h1. Use eyediagram to plot the eye diagram for w5. Indicate how the bandwidth limitations of the channel have affected the output signal. Comment specifically on how the eye diagram has been affected and how this relates to the ability of the receiver to make bit decisions, i.e., to decide whether the received pulse corresponds to a "one" or a "zero". Consider the effect of additive noise on the ability of the receiver to make bit decisions and indicate how the bandwidth limitation of the channel in combination with additive noise may degrade communication performance. If bit decisions are made by sampling the signal w5 at the pulse peak locations, what is the minimum level of additive noise that would result in a bit error?

(k). Repeat Part ?? using the impulse response h2 for the channel. How do these results compare? Can you explain this based on the relative bandwidths of the two channels?

(l). Repeat Part ?? using the impulse response h2 for the channel but use a bit rate of $f_b = 1/(2T)$. How do these results compare with Parts ?? and ??. Why?

A typical receiver for a PAM system will use a "matched filter" to determine whether the received pulse is a "zero" or a "one". A matched filter is a filter whose impulse response is given by $h(t) = p(2T - t)$, where the delay of $2T$ is introduced to make the filter causal. A matched filter has exactly the same frequency response magnitude as the pulse, so it passes the pulse and rejects any additive noise outside the band of the pulse. Specifically, if the received signal $r(t)$ contains a pulse $p(t)$ and noise $n(t)$, then at the output of the matched filter, the signal $w(t) = r(t) * p(2T - t)$ is given by

$$w(t) = \int_{-\infty}^{\infty} p(\tau)p(2T + \tau - t) + n(\tau)p(2T + \tau - t)d\tau. \tag{8.9}$$

If the output of the matched filter is sampled at $t = 2T$, then

$$w(2T) = \int_{-\infty}^{\infty} p(\tau)p(\tau) + n(\tau)p(\tau)d\tau, \tag{8.10}$$

$$= E_p + \int_{-\infty}^{\infty} n(\tau)p(\tau)d\tau, \tag{8.11}$$

where $E_p$ is the energy in the pulse $p(t)$. At the output of the matched filter, the energy of the coherently integrated pulse $E_p$ will in general be significantly larger than the incoherently integrated noise.

(m). Construct the signal y7 corresponding to N=100 randomly selected bits at a bit rate of $f_b = 1/T$. Add Gaussian noise to the transmitted signal y7 by typing yn=y7+randn(size(y7))*sigma, where sigma=0.5 is the root mean square (RMS) value of the noise. Take samples of the signal yn at the center of the pulse locations and store the result in the vector ahat. Determine whether each bit was a zero or a one by using

```
ahat=sign(yn(11:fb:length(yn))).
```

Calculate the number of bit decision errors incurred by this detection strategy. Now use a matched filter and sample the output of the matched filter at the appropriate times to determine the bit estimates. Has the number of bit errors been reduced? Repeat the experiment with `sigma=1` and `sigma=1.25`.

## 8.5 Frequency Modulation

While both amplitude and frequency modulation are used in practice for the transmission of signals, frequency modulation has a number of significant advantages over amplitude modulation. First, frequency modulation allows for more efficient transmission systems, i.e., systems which operate at peak power. Secondly, transmission systems based upon frequency modulation are more robust to additive noise disturbances during transmission. You might have already noticed that FM radio stations usually have less noise than their AM counterparts. However, this superior performance is generally at the expense of extra bandwidth.

For sinusoidal frequency modulation, the message signal to be transmitted is used to vary the frequency of a sinusoidal carrier. That is, if $x(t)$ is the message signal to be transmitted, the frequency modulated signal is given by

$$y(t) = \cos\left(\omega_c t + \int_{-\infty}^{t} x(\tau)\, d\tau\right), \tag{8.12}$$

where $\omega_c$ is called the carrier frequency. The signal $x(t)$ is encoded in the instantaneous frequency of the modulated signal $y(t)$, where the instantaneous frequency $\omega_i(t)$ is defined to be the derivative of the argument of the $\cos(\cdot)$:

$$\omega_i(t) = \frac{d}{dt}\left\{\omega_c t + \int_{-\infty}^{t} x(\tau)\, d\tau\right\}, \tag{8.13}$$

$$= \omega_c + x(t). \tag{8.14}$$

At the receiving end of the communications system, the message signal $x(t)$ can in principle be completely recovered from $y(t)$.

While amplitude modulation is a linear operation, note that the operation of frequency modulation—the mapping from the message signal $x(t)$ to the modulated signal $y(t)$—is highly nonlinear. Because of this nonlinearity, analyzing frequency modulation systems requires more sophisticated mathematical tools than do AM systems. However, for certain message signals, the time- and frequency-domain tools you have learned allow you to gain significant insight into frequency modulation. Consider the message signals

$$x_1(t) = A\cos(\omega_m t), \tag{8.15}$$

$$x_2(t) = A \sum_{k=-\infty}^{\infty} g(t - kT_m), \tag{8.16}$$

where $\omega_m = 2\pi/T_m$ is the message frequency and $g(t)$ is the signal

$$g(t) = \begin{cases} 1, & 0 \le t < T_m/2, \\ -1, & T_m/2 \le t < T_m, \\ 0, & \text{otherwise}. \end{cases} \tag{8.17}$$

The signal $x_2(t)$ is called a square wave. For the remainder of this exercise, assume that the carrier frequency is fixed at $\omega_c = 2\pi$.

## Basic Problems

Assume for the following problems that the message frequency is fixed at $\omega_m = \pi/8$ and the message amplitude is fixed at A = 4.

(a). For both of the modulated signals, determine the range of instantaneous frequency on the interval $0 \le t \le T$.

(b). For `T=16` and `dt=0.05`, define the vector `t=[0:dt:T-dt]` to contain the time samples of interest. Store in `x1` and `x2` the values of $x_1(t)$ and $x_2(t)$ at the time samples in `t`.

(c). Store in the vectors `y1` and `y2` the frequency modulation of $x_1(t)$ and $x_2(t)$ at the time samples stored in `t`. As a first step, derive analytic expressions for the arguments of $\cos(\cdot)$ in Eq. (??). Plot `x1` and `y1` versus `t` on the same set of axes. Do the same for `x2` and `y2`.

(d). Use `fft` to compute the continuous-time Fourier series coefficients for each of the two frequency modulated signals. In calculating the Fourier series coefficients, assume that `T` is the period of both signals. Executing `fft(y1)` will return, to within a constant factor, the values of the Fourier series coefficients $a_k$ for $-N/2 \le k \le N/2 - 1$, where `N=T/dt` and $a_k$ is the coefficient for the complex exponential at frequency $\omega_k = 2\pi k/T$. Make a plot of the magnitude of the Fourier series coefficients versus the frequency samples $\omega_k$. Remember that you must use `fftshift` to reorder the vectors returned by `fft`.

(e). Verify that the energy in the Fourier series coefficients is centered about $\omega = \pm\omega_c$. For each modulated signal, how well does the range of instantaneous frequency determined in Part ?? predict the spread of the Fourier series coefficients about the carrier frequency?

(f). Define the bandwidth of the modulated signals to be the width of the frequency intervals centered about $\pm\omega_c$ which contain most of the energy in $y(t)$. Do you expect the bandwidth of the two modulated signals to be a function of A? Of $\omega_m$? Explain.

## Intermediate Problems

For the message signal $x_1(t)$ in Eq. (??), the frequency modulated signal can be expanded as

$$y(t) = \cos(\omega_c t)\cos\big(m\sin(\omega_m t)\big) - \sin(\omega_c t)\sin\big(m\sin(\omega_m t)\big), \tag{8.18}$$

where $m = A/\omega_m$ is known as the modulation index. If $m$ is small, then $\cos(m\sin(\omega_m t)) \approx 1$ and $\sin(m\sin(\omega_m t)) \approx m\sin(\omega_m t)$. Substituting these approximations into Eq. (??) yields the narrowband approximation of the modulated signal. Note that the modulation index, and hence the accuracy of the narrowband approximation, depends upon both the amplitude and the frequency of the message signal.

(g). Assume $\omega_m = \pi/4$ and $A = 0.2$. What is the value of the modulation index? Sketch the Fourier transform of the narrowband approximation to $y(t)$. According to your sketch, what is the bandwidth of the modulated signal, where bandwidth is defined in Part ???

(h). Store in `xx1` and `yy1` the values of the message signal and the modulated signal evaluated at the time samples in `t`. Use Eq. (??), not the narrowband approximation, to calculate the values of $y(t)$. Plot `xx1` and `yy1` on the same set of axes.

(i). Use **`fft`** to calculate the Fourier series coefficients of the modulated signal, and plot the magnitude of these coefficients using **`stem`**. Again, when calculating the Fourier series coefficients, use `T` as the period of the signal (and hence the interval of "integration"). Appropriately label the frequency axis. How does this plot agree with the Fourier series coefficients predicted by the narrowband approximation? What is the distance from $\omega = \omega_c$ to the first frequency sample for which the Fourier series coefficient is nonzero?

(j). Repeat Part ?? for $\omega_m = \pi/2$ and $A = 0.2$. What is the value of the modulation index $m$? Based upon your plot of the Fourier series coefficients, how accurate is the narrowband approximation? Define the bandwidth of the modulated signal as twice the distance (along the frequency axis) from $\omega_c$ to the last frequency value for which $|a_k|$ is within 5% of the magnitude of the Fourier series coefficient at $\omega_c$. Did the bandwidth change as the value of $\omega_m$ increased from $\pi/4$ to $\pi/2$? What does this imply about using the range of instantaneous frequency as a measure of the bandwidth of a frequency modulated signal?

(k). Repeat Part ?? for $\omega_m = \pi/4$ and $A = 1$. What is the value of the modulation index $m$? Does the distance between the nonzero Fourier series coefficients change as the value of $A$ changes? Explain. Are these locations a function of $\omega_m$?

Chapter 9

# The Laplace Transform

The Laplace transform of a signal $x(t)$,

$$X(s) = \int_{-\infty}^{+\infty} x(t)e^{-st}dt, \tag{9.1}$$

is a generalization of the continuous-time Fourier transform that is useful for studying continuous-time signals and systems. Note that when $s = j\omega$, i.e., $s$ is purely imaginary, the Laplace transform reduces to the continuous-time Fourier transform. However, many signals which do not have Fourier transforms do have Laplace transforms, making the Laplace transform a useful tool for the analysis of linear time-invariant systems. For a large class of signals, the Laplace transform can be represented as a ratio of polynomials in $s$, i.e.,

$$X(s) = \frac{N(s)}{D(s)}, \tag{9.2}$$

where $N(s)$ and $D(s)$ are called the numerator and denominator polynomials, respectively. Transforms that can be represented as a ratio of polynomials, called rational transforms, arise as the system functions for LTI systems which satisfy linear constant-coefficient differential equations. Rational transforms are completely determined, up to a scale factor, by the roots of the polynomials $N(s)$ and $D(s)$, known as zeros and poles, respectively. Because these roots play an important role in the study of LTI systems, it is convenient to display them pictorially in a pole-zero diagram. In this chapter you will explore some of the properties of LTI systems in the frequency domain using Laplace transforms. Specifically, Tutorial ?? illustrates how to make pole-zero diagrams in MATLAB. Exercise ?? explores some of the relationships between the locations of the poles and the frequency response of second-order systems. In Exercise ??, the Butterworth family of frequency-selective filters is examined. The relationship between the magnitude of a system function and the locations of its poles and zeros is explored in Exercise ??. Finally, Exercise ?? demonstrates how noncausal systems with rational system functions can be implemented using `lsim`.

## 9.1 Tutorial: Making Continuous-Time Pole-Zero Diagrams

In this tutorial you will learn how to display the poles and zeros of a rational system function $H(s)$ in a pole-zero diagram. The poles and zeros of a rational system function can be computed using the function `roots`. For example, for the LTI system with system function

$$H(s) = \frac{s-1}{s^2+3s+2}, \tag{9.3}$$

the poles and zeros can be computed by executing

```
>> b = [1 -1];
>> a = [1 3 2];
>> zs = roots(b)
zs =
     1
>> ps = roots(a)
ps =
    -2
    -1
```

A simple pole-zero plot can be made by placing an 'x' at each pole location and an 'o' at each zero location in the complex $s$-plane, i.e.,

```
>> plot(real(zs),imag(zs),'o');
>> hold on
>> plot(real(ps),imag(ps),'x');
>> grid
>> axis([-3 3 -3 3]);
```

The function `grid` places a grid on the plot and `axis` sets the range of the axes on the plot.

(a). Each of the following system functions corresponds to a stable LTI system. Use `roots` to find the poles and zeros of each system function and make an appropriately labeled pole-zero diagram using `plot` as shown above.

(i) $H(s) = \dfrac{s+5}{s^2+2s+3}$

(ii) $H(s) = \dfrac{2s^2+5s+12}{s^2+2s+10}$

(iii) $H(s) = \dfrac{2s^2+5s+12}{(s^2+2s+10)(s+2)}$

Several different signals can have the same rational expression for their Laplace transform while having different regions of convergence. For example, the causal and anticausal LTI systems with impulse responses

$$h_c(t) = e^{-\alpha t}u(t), \qquad h_{ac}(t) = -e^{-\alpha t}u(-t),$$

have a rational system function with the same numerator and denominator polynomials,

$$H_c(s) = \frac{1}{s+\alpha}, \quad \Re e(s) > -\alpha,$$
$$H_{ac}(s) = \frac{1}{s+\alpha}, \quad \Re e(s) < -\alpha.$$

However, they have different system functions, since their regions of convergence are different.

(b). For each of the rational expressions in Part ??, determine the ROC corresponding to the stable system.

(c). For the causal LTI system whose input and output satisfy the differential equation

$$\frac{dy(t)}{dt} - 3y(t) = \frac{d^2x(t)}{dt^2} + 2\frac{dx(t)}{dt} + 5x(t),$$

find the poles and zeros of the system and make an appropriately labeled pole-zero diagram.

### Intermediate Problem

For the exercises in this chapter, you will need to use the function `plotpz`, which is contained in the Computer Explorations Toolbox. The M-file for `plotpz` is also listed below for convenience. The function `plotpz` plots a pole-zero diagram for the LTI system whose numerator and denominator polynomials have the coefficients in the vectors `b` and `a`, respectively. The function will return the values of the poles and zeros in addition to making the plot. An optional argument, `ROC`, can be used to indicate the region of convergence on the diagram. By selecting `ROC` to be a point within the region of convergence of the system, `plotpz` will appropriately label the region of convergence of the system. For example, try executing

```
>> b = [1 -1];
>> a = [1 3 1];
>> [ps,zs]=plotpz(b,a,1);
>> [ps,zs]=plotpz(b,a,-2);
```

(d). Explain how `plotpz` can determine the region of convergence of a rational transform from knowledge of a single point within the ROC.

```
function [ps,zs]=plotpz(b,a,ROC)
% [ps,zs]=plotpz(b,a,ROC)
% Plots the pole-zero diagram for the LTI system with H(s)=b(s)/a(s)
% Optional argument ROC defines one point in the ROC

ps=roots(a); % determine poles
```

```
zs=roots(b); % determine zeros
ps=ps(:); % make into column vector
zs=zs(:); % make into column vector

MaxI=max(abs(imag([ps; zs; j]))); % Determine size of diagram
MaxR=max(abs(real([ps; zs; 1]))); %
plot(1.5*[-MaxR MaxR],[0 0],'w')      % Plot the real axis
hold on
text(1.5*MaxR,0,' Re')
plot([0 0],1.5*[-MaxI MaxI],'w')      % Plot the imag axis
text(0,1.5*MaxI,' Im')
plot(real(zs),imag(zs),'ro')    % Plot zeros
plot(real(ps),imag(ps),'yx')    % Plot poles

if nargin>2,                          % ROC optional
  if(any(real(ps)<ROC))             % Any poles to the left?
    lpole=max(real(ps(real(ps)<ROC)));
    plot([lpole lpole],1.5*[-MaxI MaxI],'w--')
  end
  if(any(real(ps)>ROC))             % Any poles to the right?
    rpole=min(real(ps(real(ps)>ROC)));
    plot([rpole rpole],1.5*[-MaxI MaxI],'w--')
  end
  text(ROC,-1.25*MaxI,'ROC')            % Label the ROC
end

axis('equal');                        % Make square aspect ratio
grid
hold off
```

## 9.2 Pole Locations for Second-Order Systems

In these problems, you will examine the pole locations for second-order systems of the form

$$H(s) = \frac{\omega_n^2}{s^2 + 2\zeta\omega_n s + \omega_n^2}. \tag{9.4}$$

The values of the damping ratio $\zeta$ and undamped natural frequency $\omega_n$ specify the locations of the poles, and consequently the behavior of this system. Exercise **??** explored the relationship between the time-domain and the frequency-domain behavior of this system. In this exercise, you will see how the locations of the poles affect the frequency response.

### Basic Problems

In these problems, you will examine the pole locations and frequency responses for four different choices of $\zeta$ while $\omega_n$ remains fixed at 1.

(a). Define $H_1(s)$ through $H_4(s)$ to be the system functions that result from fixing $\omega_n = 1$ in Eq. (??) while $\zeta$ is 0, 1/4, 1, and 2, respectively. Define `a1` through `a4` to be the coefficient vectors for the denominators of $H_1(s)$ through $H_4(s)$. Find and plot the locations of the poles for each of these systems.

(b). Define `omega=[-5:0.1:5]` to be the frequencies at which you will compute the frequency responses of the four systems. Use `freqs` to compute and plot $|H(j\omega)|$ for each of the four systems you defined in Part ??. How are the frequency responses for $\zeta < 1$ qualitatively different from those for $\zeta \geq 1$? Can you explain how the pole locations for the systems cause this difference? Also, can you argue geometrically why $H(j\omega)|_{\omega=0}$ is the same for all four systems?

## Intermediate Problems

In these problems, you will trace the locations of the poles as you vary $\zeta$ and $\omega_n$, and see how varying these parameters affects the frequency response of the system.

(c). First, you will vary $\zeta$ in the range $0 \leq \zeta \leq 10$ while holding $\omega_n = 1$. Define `zetarange=[0 logspace(-1,1,99)]` to get 100 logarithmically spaced points in $0 \leq \zeta \leq 10$. Define `azeta` to be a $3 \times 100$ matrix where each column is the denominator coefficients for $H(s)$ when $\zeta$ has the value in the corresponding column of `zetarange`. Define `zetapoles` to be a $2 \times 100$ matrix where each column is the roots of the corresponding column of `azeta`. On a single figure, plot the real versus imaginary parts for each row of `zetapoles` and describe the loci they trace. On your plot, indicate the following points: $\zeta = 0$, 1/4, 1, and 2. In order to get a square aspect ratio with equal length axes in your plot, you can type

```
>> axis('equal');
>> axis([-4 0 -2 2]);
```

Describe qualitatively how you expect the frequency response to change as $\zeta$ goes from 0 to 1, and then from 1 to 10.

(d). In this problem, you will hold $\zeta = 1/4$ and examine the effect of increasing $\omega_n$ from 0 to 10. Define `omegarange=[0 logspace(-1,1,99)]` to get 100 logarithmically spaced points in the region of interest. Define `aomega` and `omegapoles` analogously to the way you defined `azeta` and `zetapoles` in Part ??. On a single figure, plot the real versus imaginary parts for each column of `omegapoles`. How would you expect changing $\omega_n$ to change the frequency response $H(j\omega)$? Use `freqs` to evaluate the frequency response when $\omega_n = 2$ and $\zeta = 1/4$, and plot the magnitude of this frequency response. Compare this with the plot you made in Part ?? for $\omega_n = 1$ and $\zeta = 1/4$. How are they different? Does this match what you expected from your plot of the loci traced by `omegapoles`?

### Advanced Problem

(e). There is no reason that $\zeta$ in Eq. (??) needs to be positive. Repeat Part ?? for $\zeta$ between $-10$ and 0. When $\zeta$ is negative, can the system described by $H(s)$ be both causal and stable? Also, plot the frequency response magnitude for the system when $\zeta = -1/4$ and $\omega_n = 1$ using `freqs`. Is the system with the frequency response computed by `freqs` causal? Also explain any similarities or differences between this plot and the frequency response magnitude plotted in Part ?? for $\zeta = 1/4$.

## 9.3 Butterworth Filters

Butterworth filters are a widely used class of continuous-time LTI systems for frequency-selective filtering. The simple analytic form of their frequency response makes Butterworth filters attractive from an engineering standpoint. In this exercise, you will use the Laplace transform to design and analyze this class of filters in the frequency domain.
An $N$th-order Butterworth lowpass filter has a frequency response whose magnitude satisfies

$$|B(j\omega)|^2 = \frac{1}{1 + (j\omega/j\omega_c)^{2N}}. \tag{9.5}$$

For a Butterworth filter with a real-valued impulse response, $b(t)$, the system function satisfies

$$B(s)B(-s) = \frac{1}{1 + (s/j\omega_c)^{2N}}. \tag{9.6}$$

### Basic Problems

(a). Determine analytically the locations of the $2N$ poles of $B(s)B(-s)$.

(b). Note that if there is a pole at $s = s_p$, then there is also a pole at $s = -s_p$. Analytically determine the location of the pole for the first-order ($N = 1$) Butterworth filter that is causal and stable. By specifying the pole location of the filter, you have only determined $B(s)$ to within an amplitude scale factor. Choose the scale factor such that the gain of the filter at $s = 0$ is unity, i.e., $B(0) = 1$.

(c). For $\omega_c = 10\pi$ and $N = 1$, use `freqs` to compute the frequency response of the first-order Butterworth filter from Part ?? at the frequencies in `w=linspace(0,1000)`. Make a clearly labeled plot of the magnitude of the response at these frequencies.

(d). For $N = 3$, determine the locations of the six poles of $B(s)B(-s)$ and store them in the vector `sp`. Plot all six poles by placing an 'x' at each pole location in the complex $s$-plane. You may use `plotpz` described in Tutorial ??.

(e). Store in the vector `csp` the three poles that correspond to the Butterworth filter which is both causal and stable. Given these three pole locations, the denominator of the system function for the filter is of the form $D(s) = (s - s_1)(s - s_2)(s - s_3)$. Use the function `poly` to construct the vector `a` of polynomial coefficients for the denominator of $B(s)$ from `csp`. What is the numerator polynomial for $B(s)$ so that the lowpass filter has unity gain at $\omega = 0$, i.e., $B(j0) = 1$? Store this in the variable `b`.

(f). Use `freqs` to compute and plot the frequency response of the third-order Butterworth filter at the frequencies in `w` and store the result in `B3`. Also use `bode(b,a)` to make a Bode plot for the system. Make sure that the filter has unity gain at $\omega = 0$.

(g). Use the function `butter(N,wc,'s')` to verify that your filter is properly specified in `b` and `a`.

(h). Store in the vectors `b2` and `a2` the coefficients of $B_2(s)$, the stable, anticausal third-order Butterworth filter. Use `bode` to plot the magnitude and phase of $B_2(s)$. How could you have expected the result? Explain any similarities or differences between the plots of $B_2(s)$ and $B(s)$.

## Intermediate Problems

Consider the design of a Butterworth lowpass filter that must meet the following specifications:

- The gain of the filter $|B(j\omega)|$ should be within $\pm 0.01$ of unity in the passband, i.e., $0.99 \leq |B(j\omega)| \leq 1.01$ for $0 \leq \omega \leq 10$.
- The gain of the filter $|B(j\omega)|$ should be no greater than 0.001 in the stopband, i.e., $|B(j\omega)| \leq 0.001$ for $\omega \geq 50$.

(i). Analytically determine values for the filter order $N$ and the cutoff frequency $\omega_c$ that will meet this specification. You would like to find the lowest order $N$ that can meet these specifications, since lower order filters are easier and cheaper to implement.

(j). For the values of $N$ and $\omega_c$ you determined in Part ??, determine the numerator and denominator polynomials for the filter and store your results in the vectors `b` and `a`, respectively.

(k). Use `freqs` to compute the frequency response at the frequencies in `w=linspace(0,60)` and store your results in `B4`. Make an appropriately labeled plot of the magnitude of `B4` versus `w`. Also use `freqs` to determine the frequency response magnitude explicitly at $\omega = 0, 10$, and 50 to verify that your filter meets the specifications.

# 9.4 Surface Plots of Laplace Transforms

In this exercise, you will visually explore the surfaces defined by rational Laplace transforms and the relationship between these surfaces and the continuous-time Fourier transform. MATLAB provides helpful facilities for evaluating rational Laplace transforms on a finite region of the $s$-plane. By examining these surfaces, you can begin to develop intuition about how locations of poles and zeros affect the system function and frequency response of LTI systems.

## Basic Problems

In these problems you will learn to compute and plot the surface defined by the magnitude of a rational system function. You will also see how the magnitude of the frequency response of the system is determined by this surface.

(a). Define vectors `b1` and `a1` to be the coefficient vectors representing the numerator and denominator polynomials of the system function

$$H_1(s) = \frac{4}{s^2 + 2s + 17}, \qquad \Re e(s) > -1. \tag{9.7}$$

Using `roots`, find and plot the poles of this system as demonstrated in Tutorial **??**.

(b). Based on your pole-zero plot from Part **??**, make a rough sketch of what you expect $|H_1(j\omega)|$ to look like. Specifically, at which frequencies do you expect the magnitude of the frequency response to have peaks? Use `freqs` to compute the continuous-time Fourier transform $H_1(j\omega)$ at the frequencies in `omega=[-10:0.5:10]`, and plot $|H_1(j\omega)|$. How does this compare with your sketch?

(c). Since `freqs` only calculates $H(s)$ on the $j\omega$ axis, you will need to explicitly calculate $H(s)$ at each value of $s$. In order to compute $H_1(s)$ over a region of the $s$-plane, you must first define a matrix containing the values of $s$ where you want to evaluate $H_1(s)$. Remember the complex values of $s$ can be represented as $\sigma + j\omega$. The values of $\omega$ will be those defined in `omega` in Part **??**. For $\sigma$, define `sigma=-1+(1/8)*(1:32)` to get 32 samples in $-1 < \sigma \leq 3$. You can use `meshgrid` to define the values of $s$ as follows:

```
>> [sigmagrid,omegagrid] = meshgrid(sigma,omega);
>> sgrid = sigmagrid+j*omegagrid;
```

The matrix `sgrid` contains a sampling of the region $-1 < \sigma \leq 3$, $-10 \leq \omega \leq 10$. You can then evaluate $H_1(s)$ at each point in `sgrid` and plot the surface defined by $|H_1(s)|$ by typing

```
>> H1 = polyval(b1,sgrid)./polyval(a1,sgrid);
>> mesh(sigma,omega,abs(H1));
```

You can use the `view` command to rotate the surface and view it from any angle. The default view is at an elevation of 30 degrees and an azimuth of -37.5 degrees. Typing `view([-80 30])` will let you see the surface from just behind the line at $\Re e(s) = -1$, while `view([0 30])` lets you look straight along the $j\omega$-axis. Examine the surface from several different angles. You may find it helpful to visualize how the poles affect the frequency response by highlighting the contour of the surface along the $j\omega$-axis. You can do this by typing

```
>> hold on;
>> plot3(zeros(1,41),omega,abs(H1(:,8))+0.05,'c')
>> hold off;
```

The offset of 0.05 is necessary to make sure that the line you plot is visible just above the surface.

(d). Consider a different system function,

$$H_2(s) = \frac{4}{s^2 + s + 16.25}, \qquad \Re e(s) > -\frac{1}{2}. \tag{9.8}$$

Find and plot the poles of this system function using `plotpz`. Based on your pole-zero plot and the surface you plotted in Part ??, how would you expect $|H_2(j\omega)|$ to differ from $|H_1(j\omega)|$? Define `H2` to be the frequency response of the new system evaluated using `freqs`. Plot the magnitude of this frequency response to confirm your answer.

## Intermediate Problems

In these problems you will examine the effect on the frequency response of adding zeros to the system function. In addition, you will explore the system function over a different region of the $s$-plane.

(e). Consider the system function

$$H_3(s) = \frac{0.25(s^2 + 1)}{s^2 + 2s + 17}, \qquad \Re e(s) > -1. \tag{9.9}$$

Generate a pole-zero plot for this system function. Based on the pole-zero plot, indicate how you expect $|H_3(j\omega)|$ to differ from $|H_1(j\omega)|$, and make a sketch of $|H_3(j\omega)|$. Use `freqs` to verify your sketch. Use the technique from Part ?? to generate a `mesh` plot of $|H_3(s)|$ over the portion of the $s$-plane in `sgrid`. Compare this surface with the one you found in Part ??. How has the inclusion of the additional zeros affected the Laplace transform?

(f). Define `s2grid=sgrid-2` to get sampling of the region $-3 < \sigma \leq 1$, $-10 \leq \omega \leq 10$. Evaluate the rational expression $H_1(s)$ from Part ?? over this new region and store the result in `newH1`. Plot the surface given by the magnitude of `newH1`. Note that you will have to adjust the `sigma` argument of `mesh` to reflect the shifted real axis of the region you have evaluated, i.e., `mesh(sigma-2,omega,abs(newH1))`. Is the surface defined by the magnitude of the rational expression finite for $\Re e(s) < -1$? Note that the region of convergence given in Eq. (??) indicates that the system function is only defined for $\Re e(s) > -1$. What system function corresponds to the surface you have found for $\Re e(s) < -1$? Can you determine the impulse response of this system analytically? Is this system stable?

## Advanced Problems

Let $X(s)$ be the Laplace transform of the signal $x(t)$. The value of $X(s)$ along any contour parallel to the imaginary axis, i.e., $s = \sigma_0 + j\omega$ for a constant $\sigma_0$, is equal to the continuous-time Fourier transform of $e^{-\sigma_0 t}x(t)$. This can be used to compute a numerical approximation to the Laplace transform when you have samples of $x(t)$, but do not have the analytic form of $x(t)$ or $X(s)$. Consider the continuous-time signal

$$x(t) = e^{-t}\cos(4t)u(t) - (1/8)e^{-t}\sin(4t)u(t). \tag{9.10}$$

(g). Find the analytic expression for $X(s)$, the Laplace transform of $x(t)$. Generate a pole-zero plot for $X(s)$. Make a sketch of the magnitude of the continuous-time Fourier transform $|X(j\omega)|$. Also, sketch $|X(s)|$ along the contour $\Re e(s) = -1/2$.

(h). In Exercise **??** you saw that `fft` could be used to compute an approximation to samples of the continuous-time Fourier transform of a signal $x(t)$ from samples $x(nT)$. Define `x` to be 64 samples of $x(t)$ with $T = 1/8$ for $0 \leq t < 8$. Imagine that you only have these samples, and do not know the analytic form for either $x(t)$ or $X(s)$. Use `fft` to compute approximate values of $X(j\omega)$ for $\omega_k = -\pi/T + 2\pi k/64T$ for $0 \leq k \leq 63$ and plot $|X(j\omega)|$. Does this agree with your sketch in Part **??**?

(i). Compute an approximation to $X(s)$ for $s = -1/2 + j\omega_k$ using `fft` and store the result in `Xhalf`. Plot $|X(s)|$ evaluated on the line $\Re e(s) = -1/2$. Is that what you would have expected to see from your pole-zero plot in Part **??**? Can you explain any discrepancies between the result you found in `Xhalf` and your sketch in Part **??**?

## 9.5 Implementing Noncausal Continuous-Time Filters

The functions `impulse` and `lsim` can be used to simulate the impulse response and output, respectively, of LTI systems whose input and output satisfy linear constant-coefficient differential equations. These functions assume that the LTI system associated with the differential equation

$$\sum_{k=0}^{K} a_k \frac{d^k y(t)}{dt^k} = \sum_{m=0}^{M} b_m \frac{d^m x(t)}{dt^m} \tag{9.11}$$

is causal, i.e., the impulse response $h(t)$ is nonzero only for $t \geq 0$. However, there are multiple LTI systems which satisfy any linear constant-coefficient differential equation, and which system is chosen depends upon the auxiliary conditions used to constrain the system output $y(t)$. For instance, initial rest conditions lead to a causal system, while final rest conditions lead to an anticausal system[1]. (An anticausal system has an impulse response $h(t)$ which is nonzero only for $t \leq 0$.) Various sets of auxiliary conditions lead to other LTI systems which satisfy Eq. (**??**).
In many applications, one is interested only in systems which are stable. The causal system associated with Eq. (**??**) is not necessarily stable. For instance, if the system function $H(s)$ has a pole in the right-half plane, then the causal system will be unstable. In this exercise, you will learn how to use `lsim` to calculate the response of the stable LTI system whose input and output satisfy Eq. (**??**).

### Basic Problems

Using `lsim` to calculate the response of the anticausal LTI system satisfying Eq. (**??**) requires a number of steps. The first step is to find a causal system whose impulse response $h_c(t)$ is a time-reversal of the impulse response for the desired anticausal system $h_{ac}(t)$, i.e.,

[1] Initial (final) rest conditions specify that $y(t) = 0$ for $t \leq \tau$ $(t \geq \tau)$ when $x(t) = 0$ for $t \leq \tau$ $(t \geq \tau)$.

$h_c(t) = h_{ac}(-t)$. This causal system can be simulated using `lsim`. A differential equation for the causal system can be derived by substituting $-t$ for $t$ in Eq. (??), which yields

$$\sum_{k=0}^{K}(-1)^k a_k \frac{d^k y(-t)}{dt^k} = \sum_{m=0}^{M}(-1)^m b_m \frac{d^m x(-t)}{dt^m}\,. \tag{9.12}$$

The input to this system is $r(t) = x(-t)$ and the output is $w(t) = y(-t)$. The auxiliary conditions for this causal system are initial rest conditions. The next step is to obtain the input $r(t)$ to the causal system by time-reversing the input $x(t)$ to the anticausal system. The third step is to use `lsim` to simulate the response $w(t)$ of the causal LTI system satisfying Eq. (??) when the input is $r(t)$. The final step is to time-reverse the simulated response $w(t)$ to obtain $y(t)$ for the original anticausal system. These steps are carried out in the following problems.

(a). If $H_{ac}(s)$ is the system function associated with Eq. (??) and $H_c(s)$ is the system function associated with Eq. (??), how are the poles of these two system functions related? How is $H_c(s)$ related to $H_{ac}(s)$?

For the following problems, consider the differential equation

$$\frac{dy(t)}{dt} + 2y(t) = x(t)\,. \tag{9.13}$$

(b). Determine the system function $H(s)$ associated with Eq. (??) and all of the possible regions of convergence of $H(s)$. For each region of convergence, determine the impulse response of the corresponding LTI system.

(c). For each region of convergence determined in Part ??, what is the corresponding auxiliary condition for the differential equation.

(d). Define `a` and `b` to contain the coefficients of the denominator and numerator polynomials of $H(s)$. For the causal system which has system function $H(s)$, use `impulse` to verify the analytic expression for the impulse response. Store in the vector `h` the values of the impulse response at the times in `t=[-5:0.01:5]` returned by `impulse`. Plot `h` versus `t`. Note that since `hs=impulse(b,a,ts)` returns samples of the response to $\delta(t-$`ts(1)`$)$, you will have to select the appropriate time samples to input to `impulse` and append the result to an appropriate number of zeros.

(e). Repeat Part ?? for the anticausal system. Remember that `impulse(b,a,ts)` assumes that the coefficients in `a` and `b` correspond to a causal system. You will need to define a new set of coefficients for a time-reversed system (see Eq. (??)) and then appropriately flip the impulse response computed by `impulse` (since the output of Eq. (??) is time-reversed).

(f). Analytically calculate the output of the anticausal LTI system satisfying Eq. (??) when the input is $x(t) = e^{5t/2}u(-t)$.

(g). Use `lsim` to verify the output of the anticausal system derived in Part ?? at the time samples `t`. Like `impulse`, the function `lsim(b,a,x,ts)` also assumes that the vectors `a` and `b` correspond to a causal system, so you must work with the time-reversed differential equation. The coefficients of this differential equation should already have been computed in Part ??. Note that the input to the time-reversed system must be time-reversed, i.e., $r(t) = x(-t)$.

## Intermediate Problems

Now consider the third-order differential equation

$$\frac{d^3y(t)}{dt^3} + \frac{d^2y(t)}{dt^2} + 24\,\frac{dy(t)}{dt} - 26\,y(t) = \frac{d^2x(t)}{dt^2} + 7\,\frac{dx(t)}{dt} + 21\,x(t)\,. \tag{9.14}$$

(h). Determine the system function $H(s)$ associated with Eq. (??) and plot the poles and zeros of this system function as shown in Tutorial ??. Determine all of the possible regions of convergence. For which region of convergence is the system stable?

(i). Use `residue` to determine the partial fraction expansion of $H(s)$. For each region of convergence determined in Part ??, analytically determine the associated impulse response.

(j). For the causal system associated with Eq. (??), use `impulse` to verify the corresponding impulse response computed in Part ??. Use the vector `t` for the time samples. What are the auxiliary conditions on $y(t)$ associated with this causal system?

(k). For the anticausal system associated with Eq. (??), use `impulse` to verify the corresponding impulse response computed in Part ??. Use the vector `t` for the time samples. Remember that you must first determine the coefficients of the time-reversed differential equation, and then derive the anticausal impulse response of Eq. (??) from the causal impulse response of the time-reversed equation. What are the auxiliary conditions on $y(t)$ associated with this anticausal system?

## Advanced Problems

The following problems consider the numerical implementation of the stable LTI system whose inputs and outputs satisfy Eq. (??). The impulse response $h_s(t)$ of this system, as calculated in Part ??, should be nonzero for all time $t$. This system is called noncausal since it is neither causal nor anticausal, and a simple time-reversal of the differential equation will not suffice for implementation. Instead, the differential equation must be decomposed into a causal and an anticausal component, each of which are computed separately. This is known as a parallel realization of the system.

(l). A parallel realization an LTI system is easily visualized by decomposing the system function as $H(s) = H_1(s) + H_2(s)$, so that the output $y(t)$ can be computed by separately computing the response to $H_1(s)$ and $H_2(s)$. Note that the regions of convergence of $H_1(s)$ and $H_2(s)$ must be properly specified. Given the partial fraction expansion you found in Part ??, determine $H_1(s)$ and $H_2(s)$ for the stable system such that $h_1(t)$ is causal and $h_2(t)$ anticausal. Make sure to specify the regions of convergence of both $H_1(s)$ and $H_2(s)$.

(m). Determine $h_1(t)$ and $h_2(t)$, the inverse transforms of $H_1(s)$ and $H_2(s)$. You should be able to determine these impulse responses from the expression for $h_s(t)$.

(n). Define $y_1(t)$ to be the output of the system defined by $H_1(s)$ and its region of convergence. Specify the differential equation corresponding to $H_1(s)$ along with the associated auxiliary conditions on $y_1(t)$.

(o). Define $y_2(t)$ to be the output of the system defined by $H_2(s)$ and its region of convergence. Specify the differential equation corresponding to $H_2(s)$ along with the associated auxiliary conditions on $y_2(t)$.

(p). Use `lsim` to determine the response of the stable system to the input

$$x(t) = \begin{cases} 1, & -3 \leq t \leq 2, \\ 0, & \text{otherwise}, \end{cases} \tag{9.15}$$

on the interval $-10 \leq t \leq 10$ for the time samples in `t=[-10:0.01:10]`. Plot $y_1(t)$, $y_2(t)$ and $y(t)$ on the interval $-10 \leq t \leq 10$.

Chapter 10

# The z-Transform

The bilateral $z$-transform of a discrete-time signal $x[n]$,

$$X(z) = \sum_{n=-\infty}^{\infty} x[n]z^{-n}, \tag{10.1}$$

is a generalization of the discrete-time Fourier transform that is useful for studying discrete-time signals and systems. Note that for $z = e^{j\omega}$, the $z$-transform reduces to the discrete-time Fourier transform. However, the bilateral $z$-transform exists for a broader class of signals than the discrete-time Fourier transform does, and is useful for understanding the behavior of both stable and unstable systems. For a large class of signals, the $z$-transform can be represented as a ratio of polynomials in $z$, i.e.,

$$X(z) = \frac{N(z)}{D(z)}. \tag{10.2}$$

These transforms are called rational transforms and arise as the system functions of LTI systems which satisfy linear constant-coefficient differential equations. The locations of the roots of $N(z)$ and $D(z)$, known as the zeros and poles of the system, respectively, determine to within a constant multiplicative factor the behavior of LTI systems with rational transforms. Therefore, plots of the pole and zero locations can be used to analyze system properties.
Tutorial ?? demonstrates how to generate a pole-zero plot for a system with a rational system function. Exercise ?? explores how the locations of the poles and zeros affect the frequency response of a system. The effect of quantizing filter coefficients on the locations of the poles and zeros of a system, and thus on the system behavior, is examined in Exercise ??. Designing discrete-time filters by simulating continuous-time filters or transforming continuous-time filters to discrete-time filters is a common signal processing application. Exercise ?? presents one transformation method known as the Euler approximations, while Exercise ?? presents another approach known as the bilinear transformation.
Note that throughout this chapter $\Omega$ is used to denote the continuous-time frequency variable, while $\omega$ is used for the discrete-time frequency variable. This convention should help minimize confusion in the exercises that refer to both domains.

## 10.1 Tutorial: Making Discrete-Time Pole-Zero Diagrams

In this tutorial you will learn how to display the poles and zeros of a discrete-time rational system function $H(z)$ in a pole-zero diagram. The poles and zeros of a rational system function can be computed using `roots` as shown in Tutorial ??. The function `roots` requires the coefficient vector to be in descending order of the independent variable. For example, consider the LTI system with system function

$$H(z) = \frac{z^2 - z}{z^2 + 3z + 2}. \tag{10.3}$$

The poles and zeros can be computed by executing

```
>> b = [1 -1 0];
>> a = [1 3 2];
>> zs = roots(b)
zs =
     0
     1
>> ps = roots(a)
ps =
    -2
    -1
```

It is often desirable to write discrete-time system functions in terms of increasing order of $z^{-1}$. The coefficients of these polynomials are easily obtained from the linear constant-coefficient difference equation and are also in the form that `filter` or `freqz` requires. However, if the numerator and denominator polynomials do not have the same order, some poles or zeros at $z = 0$ may be overlooked. For example, Eq. (??) could be rewritten as

$$H(z) = \frac{1 - z^{-1}}{1 + 3z^{-1} + 2z^{-2}}. \tag{10.4}$$

If you were to obtain the coefficients from Eq. (??), you would get the following:

```
>> b = [1 -1];
>> a = [1 3 2];
>> zs = roots(b)
zs =
     1
>> ps = roots(a)
ps =
    -2
    -1
```

Note that the zero at $z = 0$ does not appear here. In order to find the complete set of poles and zeros when working with a system function in terms of $z^{-1}$, you must append zeros to the coefficient vector for the lower-order polynomial such that the coefficient vectors are the same length.

For the exercises in this chapter, you will need the M-file `dpzplot.m`, which is in the Computer Explorations Toolbox. For convenience, the M-file is also listed below. The function `dpzplot(b,a)` plots the poles and zeros of discrete-time systems. The inputs to `dpzplot` are in the same format as `filter`, and `dpzplot` will automatically append an appropriate number of zeros to `a` or to `b` if the numerator and denominator polynomials are not of the same order. Also, `dpzplot` will include the unit circle in the plot as well as an indication of the number of poles or zeros at the origin—if there are more than one.

(a). Use `dpzplot` to plot the poles and zeros for $H(z)$ in Eq. (??).

(b). Use `dpzplot` to plot the poles and zeros for a filter which satisfies the difference equation

$$y[n] + y[n-1] + 0.5y[n-2] = x[n].$$

(c). Use `dpzplot` to plot the poles and zeros for a filter which satisfies the difference equation

$$y[n] - 1.25y[n-1] + 0.75y[n-2] - 0.125y[n-3] = x[n] + 0.5x[n-1].$$

```
function dpzplot(b,a)
% dpzplot(b,a)
% Plots the pole-zero diagram for the discrete-time system function
% H(z)=b(z)/a(z) defined by numerator and denominator polynomials b and a.

la=length(a);
lb=length(b);
if (la>lb),
  b=[b zeros(1,la-lb)];
elseif (lb>la),
  a=[a zeros(1,lb-la)];
end
ps = roots(a);
zs = roots(b);
mx = max( abs([ps' zs' .95]) ) + .05;
clf
axis([-mx mx -mx mx]);
axis('equal');
hold on
w = [0:.01:2*pi];
plot(cos(w),sin(w),'.');
plot([-mx mx],[0 0]);
plot([0 0],[-mx mx]);
text(0.1,1.1,'Im','sc');
text(1.1,.1,'Re','sc');
plot(real(ps),imag(ps),'x');
```

```
plot(real(zs),imag(zs),'o');
numz=sum(abs(zs)==0);
nump=sum(abs(ps)==0);
if numz>1,
  text(-.1,-.1,num2str(numz));
elseif nump>1,
  text(-.1,-.1,num2str(nump));
end
hold off;
```

## 10.2 Geometric Interpretation of the Discrete-Time Frequency Response

This exercise demonstrates how the magnitude and phase of the frequency response of discrete-time systems can be computed from a geometric consideration of the locations of the poles and zeros of the system function. Recall that the system function can be factored in the form

$$H(z) = A\,\frac{\prod_{k=1}^{K}(z - z_k)}{\prod_{m=1}^{M}(z - p_m)}, \tag{10.5}$$

where the $z_k$ are the $K$ zeros and the $p_m$ are the $M$ poles. The contribution of each pole and each zero to $|H(e^{j\omega})|$ depends on the length of the vector from the pole or zero to the point $e^{j\omega}$. Taking the magnitude of Eq. (??) and evaluating it at $z = e^{j\omega}$ yields

$$|H(e^{j\omega})| = |A|\,\frac{\prod_{k=1}^{K}|e^{j\omega} - z_k|}{\prod_{m=1}^{M}|e^{j\omega} - p_m|}. \tag{10.6}$$

Thus, the overall magnitude of the frequency response is the magnitude of the constant $A$ times the product of the lengths of the zero vectors divided by the product of the lengths of the pole vectors. Similarly, the contribution of each pole or zero to the phase of the frequency response $\angle H(e^{j\omega})$ is the angle formed by the real axis and the vector between the pole or zero and the point $e^{j\omega}$. Taking the phase of Eq. (??) gives

$$\angle H(e^{j\omega}) = \angle A + \sum_{k=1}^{K}\angle(e^{j\omega} - z_k) - \sum_{m=1}^{M}\angle(e^{j\omega} - p_m). \tag{10.7}$$

From this, the total phase is the phase of the constant $A$ plus the sum of the angle contributions from the zeros minus the sum of the angle contributions from the poles. For the system with the pole-zero plot shown in Figure ??, $|H(e^{j\omega})|$ is $|A\mathbf{v}_1|/|\mathbf{v}_2|$, while $\angle H(e^{j\omega})$ is $\angle A + \phi_1 - \phi_2$. The problems below will give you practice interpreting pole-zero plots in this way to obtain the frequency response of a system.

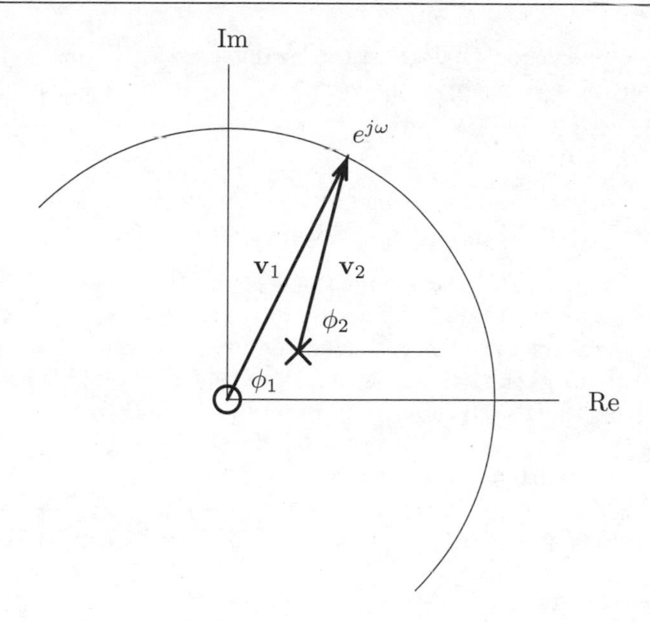


**Figure 10.1.** A geometric interpretation of the effect of poles and zeros on the discrete-time frequency response.

## Basic Problems

For these problems, you will examine a simple second-order discrete-time system whose system function is

$$H_1(z) = \frac{1}{1 - 0.9z^{-1} + 0.81z^{-2}}, \qquad |z| > 0.9. \tag{10.8}$$

(a). Define `b1` and `a1` to contain the coefficients of the numerator and denominator polynomials of $H_1(z)$ in the format required by `filter`. As described in Tutorial ??, find and plot the poles and zeros for $H_1(z)$.

(b). Define `omega=[0:511]*pi/256` and `unitcirc=exp(j*omega)` to get the 512 equally spaced points on the unit circle where you will evaluate the frequency response $H_1(e^{j\omega})$. Define `polevectors1` to be a $2 \times 512$ matrix where each row contains the complex numbers that result from subtracting one of the pole locations from the corresponding column of `unitcirc`. If `ps1` is a column vector containing the pole locations, you can do this using

```
>> polevectors1 = ones(2,1)*unitcirc-ps1*ones(1,512);
```

Using matrix and vector operations is preferable in MATLAB, because MATLAB almost always computes matrix operations faster than those constructed with `for` loops. Use `abs` and `atan2` to define `polelength1` and `poleangle1` as the magnitude and angle of each element of `polevectors1`.

(c). Define `zerovectors1` analogously to `polevectors1` so that it is the $2 \times 512$ matrix containing the vectors from the zero locations to the elements of `unitcirc`. Define `zerolength1` and `zeroangle1` to be the magnitude and the phase for these vectors, respectively.

(d). Plot `polelength1` and `zerolength1` against `omega`. Based on these plots, where do you expect $|H_1(e^{j\omega})|$ to have its maxima and minima?

(e). Use `polelength1` and `zerolength1` to compute $|H_1(e^{j\omega})|$ and store the result in `geomH1mag`. Use `poleangle1` and `zeroangle1` to compute $\angle H_1(e^{j\omega})$ and store the result in `geomH1phase`. You may find the functions `prod` and `sum` useful for defining `geomH1mag` and `geomH1phase`. Plot the geometrically derived magnitude and phase, and compare them with those you obtain by computing

```
>> H1 = freqz(b1,a1,512,'whole');
```

Was your estimate of the peak frequencies in Part **??** correct?

## Intermediate Problems

In the first set of these problems, you will examine the effect of moving one of the zeros of $H_1(z)$. Specifically, consider the system function

$$H_2(z) = \frac{1 - 0.5z^{-1}}{1 - 0.9z^{-1} + 0.81z^{-2}}, \qquad |z| > 0.9. \tag{10.9}$$

(f). Find and plot the poles and zeros of $H_2(z)$. How do you expect `polevectors2` or `zerovectors2` for this system to be different than they were for $H_1(z)$?

(g). Compute `polevectors2` and `zerovectors2` for $H_2(z)$, as well as the magnitudes and angles for all of the vectors. Plot the magnitudes and angles against `omega`. Was your prediction in Part **??** correct?

(h). Based on changes to the zeros, predict how $H_2(e^{j\omega})$ will differ from $H_1(e^{j\omega})$. Compute `H2` using `freqz` to confirm your answer.

For the problems below, you will consider the system function

$$H_3(z) = \frac{0.25 - (\sqrt{3}/2)z^{-1} + z^{-2}}{1 - (\sqrt{3}/2)z^{-1} + 0.25z^{-2}}, \qquad |z| > 0.5. \tag{10.10}$$

(i). Find and plot the poles and zeros of $H_3(z)$. How are the pole and zero locations related?

(j). Define `polevectors3` and `zerovectors3` analogously to the way you did in the Basic Problems. Define `polelength3` and `zerolength3` to be the magnitudes of these complex numbers. Plot all of these magnitudes, i.e., the magnitude of each row of `polelength3` and `zerolength3`, on the same set of axes. How are these magnitudes related? Based on this, how do you expect the frequency response magnitude $|H_3(e^{j\omega})|$ to vary with frequency? Use the lengths to compute the frequency response magnitude and store it in `geomH3mag`. Plot `geomH3mag` against `omega`.

(k). Compute H3 using `freqz` and confirm your answer from Part ??.

## 10.3 Quantization Effects in Discrete-Time Filter Structures

The function `ellip` can be used to design either discrete-time or continuous-time elliptic filters. Elliptic frequency-selective filters have a frequency response magnitude which is equiripple, i.e., the frequency response magnitude oscillates between $1 \pm \delta_1$ in the passband and between $\pm\delta_2$ in the stopband. In this set of problems, you will learn how different implementations of the eighth-order elliptic filter returned by the call

```
>> [b,a] = ellip(4,.2,40,[.41 .47]);
```

are affected by coefficient quantization. This filter has 0.2 dB ripple in the passband, $0.41\pi \le |\omega| \le 0.47\pi$, and has 40 dB of attenuation in the stopband. When discrete-time filters are implemented with quantized coefficients the resulting systems are called digital filters. Digital filters are generally implemented using fixed-point arithmetic on integer digital signal processing (DSP) chips. The coefficients of the digital filter must be quantized to the number of bits available on the DSP chip. In this exercise you will observe the effects of coefficient quantization on the frequency response of a for the different filter implementations shown in Figure ??. Specifically, you will consider three different implementations: transposed direct form (sometimes referred to as canonical form), a cascade of second-order direct form subsections, and a parallel combination of second-order direct form subsections.

### Basic Problems

(a). Store in the vectors `b` and `a` the coefficients for the eighth-order elliptic bandpass filter given above. Use `[H,w]=freqz(b,a,4096)` to compute 4096 samples of the frequency response for the filter for $0 \le \omega < \pi$. Plot the log magnitude in decibels (dB) of `H` versus `w/pi` by executing

```
>> plot(w/pi,20*log10(abs(H)));
>> axis([0 1 -80 10]);
```

(b). Use `axis` to zoom in on the passband of the filter and verify that it is equiripple in the passband. Note that the filter returned by `ellip` is scaled such that the passband oscillates between 1 and $1 - 2\delta_1$. How could you change the filter coefficients to make the frequency response oscillate between $1 \pm \Delta_1$ in the passband and still be equiripple in the stopband? Does $\Delta_1 = \delta_1$? For the following problems, use the original filter given by the coefficients in `b` and `a`.

(c). Use `filter` to compute 4096 samples of the impulse response of the filter, storing the result in the vector `h`. Plot the first 200 samples of the impulse response. What features of the impulse response indicate that this is a bandpass filter?

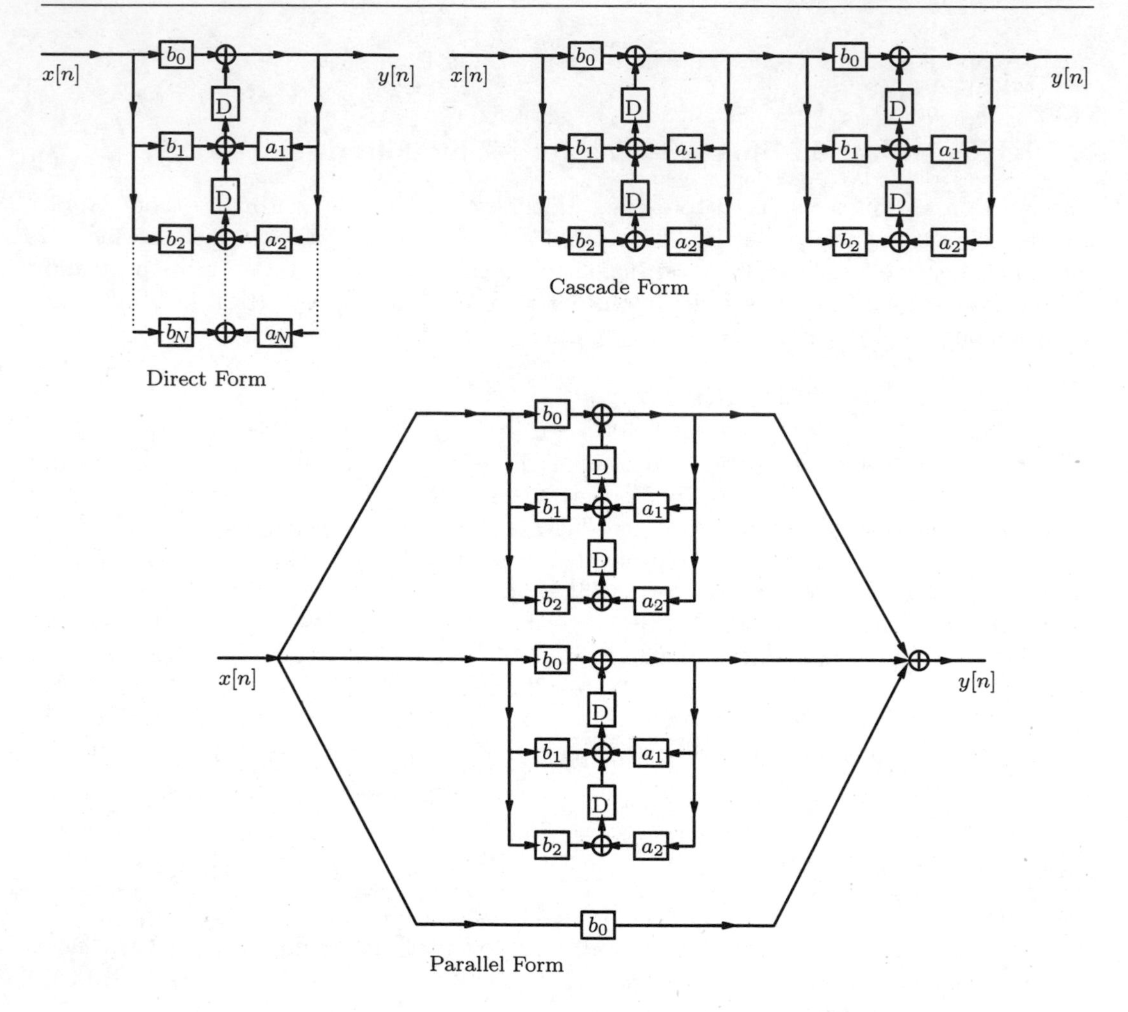


**Figure 10.2.** Three different structures for implementing discrete-time filters.

The function `y=filter(b,a,x)` uses a direct form structure—a transposed direct form II structure, to be precise—to implement the causal LTI system represented by the difference equation

```
y(n) = b(1)*x(n) + b(2)*x(n-1) + ... + b(nb+1)*x(n-nb)
                 - a(2)*y(n-1) - ... - a(na+1)*y(n-na),
```

where `a(1)` $= 1$ and `na+1` and `nb+1` are the lengths of the vectors `a` and `b`, respectively. However, when the difference equation is implemented in hardware, the direct form can be sensitive to coefficient quantization. The function `quant(x,N,M)`, which is listed here and is in the Computer Explorations Toolbox, will quantize the coefficients in the vector `x` to `N` bits, where `M` is the maximum possible amplitude of each element. The following problems examine the sensitivity of the direct form structure to coefficient quantization.

```
function qc = quant(x,N,M)
%QUANT  Q = QUANT(x,N,M) quantizes the values of x(n) into
%       2^N values.  The argument M is the value of
%       the maximum amplitude of x(n).

[mm,nn]=size(x);
qc=zeros(mm,nn);
levels = 2^(N-1);
maxlevel = 2^N-1;

for k=1:mm,
  tmp = fix((x(k,:)+M) ./ (M/levels));
  q = zeros(1,nn);
  q(tmp <= maxlevel) = tmp(tmp <= maxlevel);
  q(tmp > maxlevel) = maxlevel * ones(1,length(tmp(tmp > maxlevel)));
  q(tmp < 0) = zeros(size(tmp(tmp < 0)));
  q = (q-levels)*M/levels;
  qc(k,:)=q;
end
```

(d). Set `M=max(abs([b a]))` and use `quant` to quantize the coefficients in `b` and `a` to 16 bits storing the results in `a16` and `b16`. What is the maximum amount any of the coefficients is changed by quantization? Use `freqz` to plot the frequency response magnitude of the quantized filter as in Part ?? and zoom in on the passband as in Part ??.

(e). Use the function `dpzplot` to generate a pole-zero plot for the filter described by the coefficients `a16` and `b16`.

(f). Repeat Parts ?? and ?? quantizing `a` and `b` to 12 bits and store the results in `a12` and `b12`. Based on the pole-zero plots, is the filter whose frequency response you have plotted both causal and stable? Why or why not?

(g). Use `filter` to generate and plot 4096 samples of the impulse response of the filter described by `a12` and `b12`. Is the filter whose impulse response you have plotted stable? Why or why not?

## Intermediate Problems

In these problems, you will examine the sensitivity of a cascade of four second-order direct form subsections to coefficient quantization. You will also compare this sensitivity to that of the direct form implementation you examined in the previous problems.

(h). The MATLAB function `[bc,ac]=df2cf(b,a)`, which is listed below Part ?? and is in the Computer Explorations Toolbox, transforms the filter described by `a` and `b` into a cascade of second-order subsections. Each row of `ac` and `bc` contains the coefficients for one of the second-order subsections. Use `df2cf` with the original unquantized coefficients in `b` and `a` to create the cascade system parameters and store them in `bc` and `ac`.

(i). Use successive calls to `filter` with each row of `ac` and `bc` to generate 4096 samples of the impulse response of the cascade system; store the result in `hc`. Plot the first 200 samples of `hc` and compare this result to the plot of `h` to verify that you have implemented the filter correctly. If you have, then `max(hc-h)` should be roughly `3e-13`.

(j). Use `quant` to quantize the coefficients in `bc` and `ac` to 16 bits and store the results in `bcq16` and `acq16`. Repeat Part ?? to generate 4096 samples of the impulse response of the quantized cascade system and store the result in `hc16`.

(k). Use `freqz(hc16,1,4096)` to generate plots of the frequency response magnitude (in dB) for the cascade system and zoom in on the passband as you did in Part ??. Also use `dpzplot` to make a pole-zero plot for each of the second-order subsections described by the rows of `bcq16` and `acq16`. Are each of the causal second-order subsections stable? How do the poles and zeros of the second-order subsections correspond to the poles and zeros of the overall cascade system? How does the magnitude response of the 16-bit quantized cascade filter compare to that of the 16-bit quantized direct form implementation?

(l). Repeat Parts ??-?? quantizing `bc` and `ac` to 12 bits.

```
function [bc,ac]=df2cf(b,a)
% [bc,ac]=df2cf(b,a)
% Convert from direct form to a cascade of
% second order subsections.

N=length(b)-1;

z = roots(b);
p = roots(a);
for k=1:N/2,
    bc(k,:) = poly(z(2*k-1:2*k));
    ac(k,:) = poly(p(2*k-1:2*k));
end
bc(1,:) = b(1)*bc(1,:);
```

## Advanced Problems

In this set of problems you will consider a parallel implementation of the filter using a combination of second-order subsections. You will compare the sensitivity of this implementation to coefficient quantization with the sensitivity of the cascade and direct implementations.

(m). Use the function `[R,P,K]=residue(b,a)` to factor the system function into its partial fraction expansion. Remember that using `residue` on polynomials in $z^{-1}$ requires the coefficients of the filter to be in the reverse of the order required by `filter`, as discussed in Exercise ??. In order to implement the filter using a parallel combination of second-order subsections, you will need to combine pairs of the first-order sections from the partial fraction expansion. To simplify computation, each complex pole should be combined with its complex conjugate, so that the overall second-order subsections will have real-valued coefficients. Select four pairs of first-order terms and use `residue` to recombine these first-order terms into second-order subsections. If there is any `K` term in the partial fraction expansion corresponding to a term of the form $K\delta[n]$ in the impulse response, be sure to include `K` in one of the calls to `residue`. Store the coefficients of the resulting second-order subsections in the matrices `bp` and `ap`, with each row corresponding to one second-order system.

(n). Use repeated calls to `filter` to implement the parallel system and generate 4096 samples of the impulse response of the system, storing the result in `hp`. Plot the first 200 samples of `hp` and compare this result to the plot of `h` to verify that you have implemented the filter correctly. If you have, then `max(hp-h)` should be roughly `2e-13`.

(o). Use `quant` to quantize the coefficients in `bp` and `ap` to 16 bits and store the results in `bpq16` and `apq16`. Use `residue` on each row of `bpq16` and `apq16` to obtain the partial fraction expansion of the filter that was quantized in parallel form. Now with a single call to `residue` recombine the terms to obtain the single set of coefficients, `bp16` and `ap16`, for the overall system function. Note that although the system function is given as a single difference equation, the coefficients were quantized with the system in parallel form.

(p). Use repeated calls to `filter` to simulate the filter in quantized parallel form and generate 4096 samples of the impulse response, storing the result in `hp16`. Use `[H,w]=freqz(hp16,1,4096)` to generate plots of the frequency response magnitude in dB. Also use `freqz(bp16,ap16,4096)` to generate plots of the frequency response magnitude to verify that the system function described by `bp16` and `ap16` is indeed the same as that of the quantized parallel form system.

(q). Use `dpzplot` to make a pole-zero plot for the filter described by `bp16` and `ap16`. Is this filter both causal and stable? How does the magnitude response of the 16-bit quantized parallel filter compare to that of the 16-bit quantized cascade and direct form implementations?

(r). Repeat Parts ??–??, quantizing `bp` and `ap` to 12 bits.

## ■ 10.4 Designing Discrete-Time Filters with Euler Approximations

It is often desirable to design a discrete-time filter based on a continuous-time filter. For a continuous-time filter described by a differential equation, the continuous-time system can be transformed into a difference equation representing a discrete-time filter. Ideally, you would like this transformation to preserve the general character of the frequency response of the system. Two methods to convert differential equations to difference equations are the Euler approximations, sometimes called backward and forward differences. These approximations use the limiting definition of a derivative to replace the derivatives in a differential equation with finite differences. Specifically, the backward Euler approximation or backward difference exploits the definition

$$\frac{dx_c(t)}{dt} = \lim_{\Delta t \to 0} \frac{x_c(t) - x_c(t - \Delta t)}{\Delta t}. \tag{10.11}$$

If you define $x[n] = x_c(n\Delta t)$, you can convert the differential equation into a difference equation by replacing

$$\frac{dx_c(t)}{dt} \text{ with } \frac{x[n] - x[n-1]}{\Delta t}, \tag{10.12}$$

for some appropriately small value of $\Delta t$. While differentiating in time results in multiplication by $s$ in the continuous-time frequency domain, there will be an analogous mapping from the backwards Euler approximation to multiplication by a function $\mathrm{b}(z)$ in the discrete-time frequency domain:

$$\frac{dx_c(t)}{dt} \overset{\mathcal{L}}{\longleftrightarrow} sX_c(s)\,, \tag{10.13a}$$

$$\frac{x[n] - x[n-1]}{\Delta t} \overset{\mathcal{Z}}{\longleftrightarrow} \mathrm{b}(z)X(z)\,. \tag{10.13b}$$

Similarly, the forward Euler approximation or forward difference uses the limiting definition of the derivative,

$$\frac{dx_c(t)}{dt} = \lim_{\Delta t \to 0} \frac{x_c(t + \Delta t) - x_c(t)}{\Delta t}, \tag{10.14}$$

which leads to replacing

$$\frac{dx_c(t)}{dt} \text{ with } \frac{x[n+1] - x[n]}{\Delta t}. \tag{10.15}$$

This substitution yields the $z$-transform mapping

$$\frac{x[n+1] - x[n]}{\Delta t} \overset{\mathcal{Z}}{\longleftrightarrow} \mathrm{f}(z)X(z)\,. \tag{10.16}$$

In this exercise, you will explore the effect of converting continuous-time Butterworth lowpass and bandpass filters to discrete-time filters using the forward and backward difference approximations.

## Basic Problems

For these problems, you will work with the continuous-time lowpass Butterworth filters described by the differential equation

$$\frac{d^3y_c(t)}{dt^3} + \frac{d^2y_c(t)}{dt^2} + \frac{1}{2}\frac{dy_c(t)}{dt} + \frac{1}{8}y_c(t) = \frac{1}{8}x_c(t). \tag{10.17}$$

(a). Analytically determine the system function $H_{\rm lp}(s)$ for the filter in Eq. (??). Define `alp` and `blp` to be the coefficients of the denominator and numerator of the system function. Use `freqs` to find and plot the frequency response magnitude for the continuous-time filter at the frequencies along the $j\omega$-axis in `w=[-10:0.25:10]`. Also, plot the poles and zeros for the the system function $H_{\rm lp}(s)$ as specified in Tutorial ??.

(b). Replacing derivatives with differences in the differential equation is equivalent to replacing each occurrence of $s$ in the system function $H_c(s)$ with a function of $z$. For the backward Euler approximation in Eq. (??), this function will be $\mathrm{b}(z)$ from Eq. (??). Solve for $\mathrm{b}(z)$ using Eqs. (??) and (??) and $X_c(\mathrm{b}(z)) = X(z)$. The discrete-time system function that results from using the backward Euler approximation is then $H_{\rm b}(z) = H_c(\mathrm{b}(z))$. Similarly, for the forward Euler approximation of Eq. (??), this function will be $\mathrm{f}(z)$. Analytically derive $\mathrm{f}(z)$ using properties of the Laplace and $z$-transforms so that $H_{\rm f}(z) = H_c(\mathrm{f}(z))$.

(c). Using your expressions from Part ?? with the system function $H_{\rm lp}(s)$, analytically derive the system functions $H_{\rm blp}(z)$ and $H_{\rm flp}(z)$ for both the backward and forward differences with $\Delta t = 1/2$. For both system functions plot the magnitude of the frequency response at 1024 equally spaced points on the unit circle using `freqz` with the `'whole'` option. Which of the approximations results in a discrete-time frequency response which is a closer approximation to the original lowpass continuous-time Butterworth filter? Do the discrete-time filters attenuate high frequencies as well as $H_{\rm lp}(s)$?

(d). Use your expressions from Part ?? to predict how the poles and zeros of $H_{\rm lp}(s)$ will be mapped by the backward and forward difference approximations. Plot the poles and zeros of $H_{\rm blp}(z)$ and $H_{\rm flp}(z)$ and verify your prediction.

## Intermediate Problems

In these problems, you will convert the continuous-time Butterworth bandpass filter described by

$$\frac{d^4y_c(t)}{dt^4} + \sqrt{2}\,\frac{d^3y_c(t)}{dt^3} + 5\,\frac{d^2y_c(t)}{dt^2} + 2\sqrt{2}\,\frac{dy_c(t)}{dt} + 4\,y_c(t) = \frac{d^2x_c(t)}{dt^2} \tag{10.18}$$

into discrete-time filters using forward and backward differences with different values for $\Delta t$.

(e). Find the system function $H_{\rm bp}(s)$ for the causal system satisfying the differential equation in Eq. (??). Plot the frequency response of the continuous-time system for the frequencies in `w`. Find and plot the poles and zeros of $H_{\rm bp}(s)$.

(f). Use the expressions you found in Part ?? to convert $H_{\rm bp}(s)$ to the discrete-time system functions $H_{\rm bbp1}(z)$ and $H_{\rm fbp1}(z)$ with $\Delta t = 1$ using both the backward and forward differences, respectively. Use `freqz` to plot the magnitude of the frequency response for both of these systems. Do these systems preserve the bandpass characteristic of the original continuous-time system?

(g). Because the approximations in Eqs. (??) and (??) are based on limiting definitions of the derivative, you might expect them to get better for smaller $\Delta t$. Convert the system function $H_{\rm bp}(s)$ to the discrete-time system functions $H_{\rm bbp2}(z)$ and $H_{\rm fbp2}(z)$ using $\Delta t = 0.3$ for the Euler approximations. Find and plot the frequency responses for these new systems. Does decreasing $\Delta t$ better preserve the bandpass characteristic of the discrete-time system?

(h). An important method for evaluating the performance of a transformation from a continuous-time filter to a discrete-time filter is determining how it maps the imaginary axis ($s = j\Omega$) to the $z$-plane. Since the frequency response of a discrete-time system is the system function evaluated along the unit circle ($|z| = 1$), it is often desirable that the transformation be a one-to-one mapping from the $j\Omega$-axis to the unit circle. To examine the mappings for the Euler approximations, define the imaginary axis as the following samples of $j\Omega$:

```
>> jOmega = j*[fliplr(-logspace(-2,2,101)),0, logspace(-2,2,101)];
```

Solve your expressions for $s = \mathrm{b}(z)$ and $s = \mathrm{f}(z)$ from Part ?? to find $z$ in terms of $s$ for each approximation. Using these expressions, define `zbe` and `zfe` to be the points in the $z$-plane to which `jOmega` maps using the backward and forward Euler approximations. Generate a plot showing these contours in the $z$-plane and include the unit circle in your plot. Do either of the Euler approximations map the imaginary axis in the $s$-plane to the unit circle in the $z$-plane?

### Advanced Problem

(i). Consider the discrete-time signal

$$x[n] = \cos\left(\frac{\pi n}{5}\right) + \cos\left(\frac{\pi n}{2}\right). \tag{10.19}$$

Based on the frequency responses for the systems $H_{\rm bbp2}(z)$ and $H_{\rm fbp2}(z)$, what would you expect the output of each system to be if $x[n]$ were the input? Simulate each of the systems using `filter`. Do the outputs you get match your predictions? If not, why not?

## 10.5 Discrete-Time Butterworth Filter Design Using the Bilinear Transformation

This problem explores the design of discrete-time filters from continuous-time filters using the bilinear transformation. The bilinear transformation is a mapping that can be used to obtain a rational $z$-transform $H_{\rm d}(z)$ from a rational Laplace transform $H_{\rm c}(s)$. This mapping has some important properties, including the following:

(1). If $H_{\rm c}(s)$ is the system function for a causal and stable continuous-time system, then $H_{\rm d}(z)$ is the system function for a causal and stable discrete-time system.

(2). If $H_{\rm c}(j\Omega)$ is a piecewise constant frequency response, then $H_{\rm d}(e^{j\omega})$ will also be piecewise constant.

These two properties in combination imply that causal and stable ideal frequency-selective filters $H_c(s)$ will map to causal and stable ideal frequency-selective filters $H_d(z)$. The bilinear transformation of $H_c(s)$ is given by

$$H_d(z) = H_c(s) \quad \text{evaluated with} \quad s = \frac{1-z^{-1}}{1+z^{-1}}. \tag{10.20}$$

For the frequency response, the mapping is given by

$$H_d(e^{j\omega}) = H_c(j\Omega)|_{\Omega=\tan(\omega/2)} \,. \tag{10.21}$$

## Basic Problems

In this set of problems, you will analytically design a discrete-time lowpass Butterworth filter using the bilinear transformation. The discrete-time filter must meet the following specifications:

- passband frequency $\omega_p = 0.22\pi$,
- stopband frequency $\omega_s = 0.4\pi$,
- passband tolerance $\delta_1 = 0.3$,
- stopband tolerance $\delta_2 = 0.25$.

This implies that

$$1 - \delta_1 \leq |H_d(e^{j\omega})| \leq 1\,, \qquad 0 \leq |\omega| \leq \omega_p\,, \tag{10.22}$$

$$|H_d(e^{j\omega})| \leq \delta_2\,, \qquad \omega_s \leq |\omega| \leq \pi\,. \tag{10.23}$$

An $N$th-order continuous-time Butterworth lowpass filter has a frequency response whose magnitude satisfies

$$|H_c(j\Omega)|^2 = \frac{1}{1+(j\Omega/j\Omega_c)^{2N}}\,. \tag{10.24}$$

For a Butterworth filter with a real-valued impulse response, $b(t)$, the system function satisfies

$$H_c(s)H_c(-s) = \frac{1}{1+(s/j\Omega_c)^{2N}}. \tag{10.25}$$

See Exercise ?? for more information about continuous-time Butterworth filters. The design method that you will use is

- Map the specifications for the discrete-time filter into specifications for a continuous-time filter using the bilinear transformation.
- Design a continuous-time Butterworth filter to meet or exceed these specifications.
- Map the continuous-time filter back to discrete-time using the bilinear transformation.

The following problems guide you through this design procedure.

(a). Map the corner frequencies $\omega_p$ and $\omega_s$ to continuous-time corner frequencies $\Omega_p$ and $\Omega_s$ using Eq. (??).

(b). Using Eqs. (??)-(??) solve for the integer filter order $N$ and the cutoff frequency $\Omega_c$ that will meet this specification. You would like to find the lowest order $N$ that can meet these specifications, since lower order filters are easier and cheaper to implement. Also determine $H_\mathrm{c}(s)$, the system function for the filter.

(c). Use the function `plotpz` to make a pole-zero plot for the continuous-time filter. Also use `freqs` to make a plot of the frequency response magnitude. Be sure to check the passband and stopband edges to verify that your filter meets or exceeds the continuous-time specifications.

(d). Map the continuous-time system function $H_\mathrm{c}(s)$ to a discrete-time system function $H_\mathrm{d}(z)$ using the bilinear transformation Eq. (??) and store the resulting numerator and denominator polynomial coefficients in the vectors `a` and `b`, respectively. Use `freqz` to plot the frequency response magnitude and to verify that your filter meets the original discrete-time specifications.

(e). Use the function `dpzplot` to make a pole-zero plot for the discrete-time filter and to verify that the causal and stable continuous-time filter maps to a causal and stable discrete-time filter.

## Intermediate Problems

The function `butter` takes two arguments, $N$ and $\omega_c$, and then designs an $N$th-order discrete-time lowpass filter with a frequency response magnitude that satisfies

$$|H_\mathrm{d}(e^{j\omega})|^2 = \frac{1}{1+\left(\dfrac{\tan(\omega/2)}{\tan(\omega_c/2)}\right)^{2N}}\,.$$

The function `butter` designs the discrete-time filter from a continuous-time filter using the bilinear transformation with the following algorithm:

- Prewarp the discrete-time cutoff frequency $\omega_c$ to the corresponding continuous-time cutoff frequency $\Omega_c$.
- Design the $N$th-order continuous-time Butterworth filter with cutoff frequency $\Omega_c$.
- Perform the bilinear transformation and return the coefficients of the discrete-time filter system function.

A more useful filter design program would accept a passband edge frequency $\omega_p$, a stopband edge frequency $\omega_s$, a passband tolerance $\delta_1$, and a stopband tolerance $\delta_2$ as inputs, and then return the coefficients and filter order of a discrete-time filter meeting these specifications.

(f). Use the following design algorithm to write a function `betterbutter`:

- Prewarp the discrete-time edge frequencies to the corresponding continuous-time frequencies.
- Compute the integer filter order $N$ and cutoff frequency $\Omega_c$ such that the filter specifications are met exactly at the passband edge. The function `ceil` may be useful for this step.
- Warp $\Omega_c$ back to the discrete-time cutoff frequency $\omega_c$.
- Use the MATLAB function `butter` to design the desired filter and return the filter coefficients along with the filter order.

The first line of your function should read

```
function [b,a,N] = betterbutter(wp,ws,d1,d2)
```

(g). Use your function to design a filter which meets the specifications in the Basic Problems. Use `freqz` to make an appropriately labeled plot of the magnitude of the frequency response. Are the resulting coefficients equal to those found analytically?

(h). Use your function to design a filter to meet the following specifications:

- passband frequency $\omega_p = 0.3\pi$,
- stopband frequency $\omega_s = 0.4\pi$,
- passband tolerance $\delta_1 = 0.05$,
- stopband tolerance $\delta_2 = 0.025$.

Use `freqz` to verify that the frequency response for this filter meets the specifications. What is the order of your filter?

Chapter 11

# Feedback Systems

In the previous chapters, the focus has been on analyzing and designing systems from an input to output perspective. Specifically, you have studied how the output signal results from a given input. In many scenarios, it is desirable to use the output signal to modify the input signal in order to obtain a desired output or system property. This modification is known as feedback, since the output signal is fed back to the input of the system and combined with the original input. One example of feedback is the navigation of an airplane. While a pilot could in theory determine ahead of time the proper thrust and steering needed to guide the plane to its destination, in practice he will be unable to predict the disturbances from pressure and wind velocity variations which will perturb the plane from its desired course. Instead, he uses instruments during flight to determine the velocity and location of the plane, which can in turn be used to adjust the pilot's inputs to the plane (thrust, steering, etc.) in order to arrive at his destination. In fact, it is this very feedback loop which is the principle behind the auto pilots in common use today.
Feedback is not only useful for accounting for unknown disturbances, but it can also be used to stabilize systems which are inherently unstable. One such unstable system is the balancing of a broomstick, where feedback can be used to overcome the instability introduced by gravity. This application of feedback is considered in Exercise ??. Feedback is also preferable to many other methods of stabilization, such as compensation, since feedback stabilization is robust to uncertainties in models for physical systems. This aspect of feedback and the use of the function `rlocus` for designing stable systems are covered in Exercise ??. Another common use of feedback is to increase the bandwidth of electrical amplifiers like operational amplifiers (op-amps). While op-amps have large open-loop gains at low frequencies, the bandwidth of the amplifier is typically very small and the amplification can vary considerably as a function of frequency over the passband. As explored in Exercise ??, feedback can be used to produce an amplifier with fairly stable gain over a much larger bandwidth than that of the open-loop amplifier. The price is a proportional decrease in the value of the passband gain.

## ■ 11.1 Feedback Stabilization: Stick Balancing

A variety of mechanical systems, including spacecraft attitude control or robot arm manipulation, share the complex problem of balancing an object which is part of an inherently unstable system. This exercise considers the problem of stabilizing the unstable system that governs the balancing of an object like a stick in your hand. Since you have probably

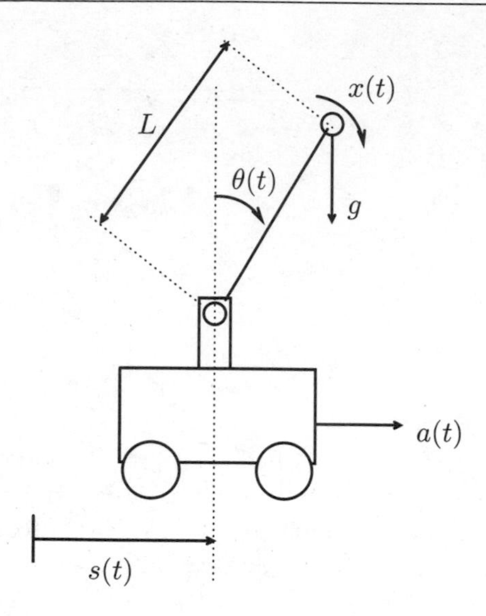


**Figure 11.1.** Stick balancing on a cart.

completed this task a number of times, you should have an idea of the difficulty of the problem as well as knowledge that it is possible. However, you should also realize that it cannot be done without feedback. Try balancing a pencil in your hand by placing the eraser end in your palm. Now try it with your eyes closed. Without the visual feedback of the location of your hand and the position of the pencil, this task is impossible. In this exercise you will examine exactly what kind of feedback is necessary for a simplified version of the stick balancing problem.

Consider the system shown in Figure ??. The cart, whose position is given by $s(t)$, can move forward or backward. The acceleration of the cart follows as $\ddot{s}(t) = a(t)$. A stick of length $L$ is attached to the cart with a hinge so that it can only move in the same direction as the cart. The position of the stick is given by the angle $\theta(t)$. Assume that all of the mass of the stick is concentrated in a ball at the end of the stick. Also shown is an angular acceleration $x(t)$ imparted on the stick by external forces, such as the wind. In order to balance the stick, the cart must be moved with an appropriate acceleration $a(t)$.

Balancing the forces on the mass along the direction perpendicular to the rod, the differential equation relating $\theta(t)$, $a(t)$, and $x(t)$ is

$$L\frac{d^2\theta(t)}{dt^2} = g\sin\big(\theta(t)\big) - a(t)\cos\big(\theta(t)\big) + L\,x(t). \tag{11.1}$$

Note that Eq. (??) is not a linear differential equation. Analysis of nonlinear systems is often difficult. However, by linearizing the equation for small $\theta(t)$, you will be able to examine the dynamics when the stick is nearly vertical (which is where you want it to be). In this case, you can make the small angle approximations

$$\sin\theta(t) \approx \theta(t), \qquad \cos\theta(t) \approx 1, \tag{11.2}$$

for $|\theta(t)| \ll \pi$. For the problems that follow, assume that $L = 1$ m and $g = 9.8$ m/s$^2$, and use the linearized equation

$$L\frac{d^2\theta(t)}{dt^2} = g\,\theta(t) - a(t) + L\,x(t). \tag{11.3}$$

## Basic Problems

(a). If the cart were stationary, $a(t) = 0$, find the system function relating the input $x(t)$ to the output $\theta(t)$. Make a pole-zero plot for this system using `plotpz` and explain from the plot why the system is unstable. This should agree with your experiences with the pencil. It would be difficult to get a pencil to balance in your hand without moving your hand at all.

(b). You will now consider stabilizing the system with proportional feedback, i.e., using an acceleration of the cart which is proportional to the angle $\theta(t)$, $a(t) = k\theta(t)$. Determine the system function for the system with proportional feedback. For the set of values, `k=linspace(0,25,10)`, plot the pole locations for the system using proportional feedback.

(c). Find a value of $k$ such that the stick location will oscillate back and forth indefinitely when $x(t) = \delta(t)$.

(d). Use `impulse` with the value of $k$ from Part ?? to simulate the impulse response of the system over `t=linspace(0,10,100)`, storing the simulated values for $\theta(t)$ in `th`. Make an appropriately labeled plot of $\theta(t)$ over this range.

## Intermediate Problems

In the Basic Problems, you should have found that the system cannot be stabilized using only proportional feedback. This means that in order to balance the stick, you would need more information about the system than simply the stick location. In the next set of problems, you will consider using proportional-plus-derivative feedback to stabilize the system, i.e., feedback of the form

$$a(t) = k_1\theta(t) + k_2\frac{d\theta(t)}{dt}.$$

(e). Analytically determine the system function for the system with proportional-plus-derivative feedback. Show that you can find values for $k_1$ and $k_2$ to stabilize the system. Analytically determine values for $k_1$ and $k_2$ so that the damping ratio ($\zeta$) of the closed-loop system is 1 and the undamped natural frequency ($\omega_n$) is 3 rad/sec. See Exercises ?? and ?? for definitions of $\zeta$ and $\omega_n$.

(f). Use `plotpz` to make a pole-zero diagram for the system using the values for $k_1$ and $k_2$ obtained in Part ??.

(g). Use `lsim` to simulate the impulse response of the system over `t=linspace(0,10,100)`, storing the simulated values for $\theta(t)$ in `th2`. Make an appropriately labeled plot of $\theta(t)$ over this range.

(h). If there were a small random disturbance $x(t)$ over $0 \leq t \leq 5$ seconds, you would like the system to be stable enough to recover, placing the stick back in the vertical position. Create the random disturbance `x=[randn(1,50) zeros(1,50)]` which is nonzero for $0 \leq t < 5$. Use `lsim` to simulate the response of the system to this input over the time samples in `t`, storing the resulting simulated values for $\theta(t)$ in `th3`. Make an appropriately labeled plot of $\theta(t)$ over this range. Does the system recover?

(i). Plot the acceleration of the cart $a(t)$ used to balance the stick in Part ??. Since the feedback is proportional-plus-derivative, you will need both $\theta(t)$ and $d\theta(t)/dt$ to create $a(t)$. Approximate $d\theta(t)/dt$ using a backwards Euler approximation (see Exercise ??). You may find the function `diff` useful. For a cart of mass $m$, what force would be required to yield $a(t)$?

## 11.2 Stabilization of Unstable Systems

There are a large number of scenarios where it is necessary to use an inherently unstable system as a component of a larger system. However, with access to both the input and output of the unstable system, feedback can be used to create an overall system, called the closed-loop system, which is stable. Consider a causal, unstable system with system function $H(s)$ and assume for this exercise that all of the system functions correspond to causal systems. A basic feedback system is illustrated in Figure ??. In this feedback system, $H(s)$ is referred to as the open-loop system, while the closed-loop system has inputs and outputs which satisfy $Y(s) = Q(s)X(s)$ for

$$Q(s) = \frac{H(s)}{1 + G(s)H(s)}. \tag{11.4}$$

Even if the open-loop system is unstable, the closed-loop system will be stable if $G(s)$ is chosen such that the poles of $Q(s)$ are in the left half of the $s$-plane. In this exercise, you will use the function `rlocus` to analyze the behavior of the open-loop system $H(s)$ and to determine parameters of the appropriate feedback system $G(s)$.

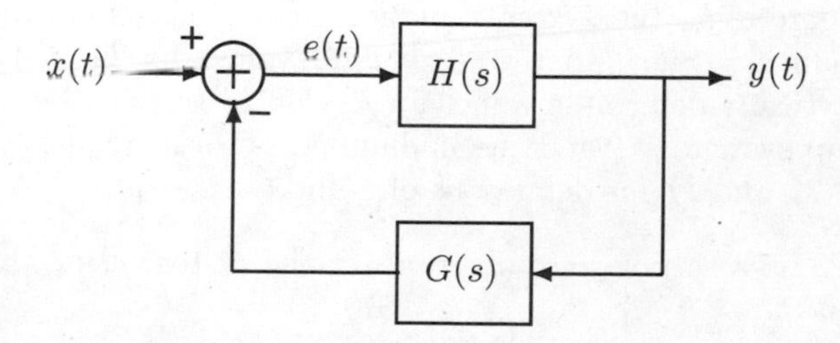


**Figure 11.2.** Using feedback to control the behavior of the open-loop system $H(s)$.

## Basic Problems

Consider the unstable system which satisfies

$$\frac{dy(t)}{dt} = 2\,y(t) + x(t)\,, \tag{11.5}$$

where $y(t)$ might describe the amount of bacteria in a petri dish over some finite time interval. The input $x(t)$ might represent the effect of the food supply or temperature on the bacterium population. Consider adding proportional feedback, $G(s) = K$, to stabilize the system.

(a). Determine the closed-loop system function $Q(s)$ as a function of $K$.

(b). Analytically determine the location of pole of the closed-loop system as a function of $K$. For what values of $K$ is the closed-loop system stable?

For the closed-loop system given by Eq. (??) and $G(s) = K$, the function `rlocus` can be used to plot the pole locations as a function of $K$. In particular, `rlocus` solves for the roots of the equation

$$1 + KH(s) = 0 \tag{11.6}$$

as a function of $K$. These roots are equal to the poles of the closed-loop system. If `b` and `a` contain the numerator and denominator polynomials of $H(s)$, then `rlocus(b,a)` plots the loci of the roots of Eq. (??) for $K \geq 0$. Note that `rlocus(-b,a)` will plot the roots for $K \leq 0$. However, the root locus plots do not allow you to determine the value of $K$ which leads to a given set of pole locations. To do this, output arguments can be supplied to `rlocus`, i.e., `[r,k]=rlocus(b,a)`, where each row of `r` contains the roots of Eq. (??) for the corresponding value of $K$ in `k`.

(c). Use `rlocus` to verify the pole locations determined in Part ??. You should create separate plots for $K \geq 0$ and $K \leq 0$. You should also use the form `[r,k]=rlocus(b,a)` to verify the values of $K$ for which the closed-loop system is stable.

Now consider the system with system function

$$H(s) = \frac{s^4 + 40s^3 + 600s^2 + 4000s + 10000}{s^4 - s^3 + 123s^2 + 50s + 2500}\,. \tag{11.7}$$

(d). For proportional feedback, $G(s) = K$, use `rlocus` to plot the location of the closed-loop poles. Use separate plots for $K \geq 0$ and $K \leq 0$.

(e). Use `rlocus` to determine a value for $K$ such that all of the closed-loop poles $p_k$ satisfy $-15 < \Re e(p_k) < 0$ and $-5 < \Im m(p_k) < 5$.

## Intermediate Problems

Another method for stabilizing an unstable system is to connect the system in series with another system whose zeros cancel the unstable poles. This method is known as compensation, and is illustrated in Figure ??. Namely, if $H(s)$ is rational and has poles at $s = p_k$

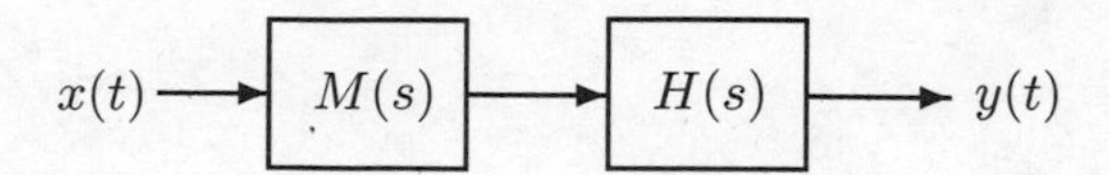


**Figure 11.3.** Using compensation to stabilize the unstable system with system function $H(s)$.

in the right-half plane, then the compensating system $M(s)$ should have zeros at $s = p_k$ to cancel the unstable poles. However, as shown in the following problems, compensation generally does not work as well in practice as feedback does. You will compare the performance of a compensated system with one using proportional feedback, i.e., $G(s) = K$. Consider the open-loop system with system function

$$H(s) = \frac{s^2 + 5\sqrt{2}s + 25}{s^2 - 1}, \tag{11.8}$$

and consider the compensator

$$M(s) = \frac{s^2 - 1}{2(s^2 + 3.6s + 12.24)}. \tag{11.9}$$

(f). Store in the vectors `a` and `b` the denominator and numerator polynomials of $H(s)$. Use `rlocus` to plot the locations of the poles for the closed-loop system with proportional feedback, $G(s) = K$. Use separate plots for $K \geq 0$ and $K \leq 0$.

(g). Use `[r,k]=rlocus(b,a)` to estimate the values of $K$ for which the closed-loop system is stable.

(h). Determine a value for $K$ such that the real parts of the closed-loop poles are within $-1.8 \pm 0.05$. For this value of $K$, store in `ak` and `bk` the denominator and numerator coefficients of the system function $Q(s)$ for the closed-loop system.

(i). For the time samples `t=[0:0.1:5]`, use `step` to compute the step response of the closed-loop system.

(j). Store in `ac` and `bc` the coefficients of the denominator and numerator of the compensated system function $H_c(s) = M(s)H(s)$. Determine the poles of the compensated system function.

(k). Use `step` to compute the step response of this system at the time samples in `t`. The step response should be very similar to the step response of the closed-loop feedback system plotted in Part ??.

In general, the exact properties of physical systems are not known. For instance, electrical resistance varies with temperature. For this reason, any system design should allow for errors in modeling. Consider the system function

$$H_\varepsilon(s) = \frac{s^2 + 5\sqrt{2}s + 25}{s^2 - \varepsilon s - (1 + \varepsilon)}, \tag{11.10}$$

where $\varepsilon$ represents uncertainty in the model parameters of the system given in Eq. (??).

(l). Determine the poles of the system function in Eq. (??) as a function of $\varepsilon$.

(m). For the remaining problems, assume $\varepsilon = 0.01$. Store in `ae` and `be` the coefficients of the denominator and numerator polynomials for the system function $H_\varepsilon(s)$.

(n). Assume $H_\varepsilon(s)$ for $\varepsilon = 0.01$ is the true open-loop system function. Using the value for $K$ computed in Part ??, store in `ake` and `bke` the coefficients of the denominator and numerator polynomial for the closed-loop system function $Q(s)$, again assuming proportional feedback $G(s) = K$. Compute the step response of this closed-loop system for the time samples in `t` and compare it with the step response computed in Part ??.

(o). Store in `ace` and `bce` the coefficients of the denominator and numerator of the compensated system function $H_{c\varepsilon}(s) = M(s)H_\varepsilon(s)$. Compute the step response of this compensated system for the time samples in `t` and compare it with the step response computed in Part ??. Explain any differences between the two step responses.

(p). Is the feedback system more robust than the compensated system to uncertainties in the open-loop system? Explain.

## 11.3 Using Feedback to Increase the Bandwidth of an Amplifier

Many systems designed as linear amplifiers use operational amplifiers (op-amps) like the LM741. The open-loop gains for op-amps generally exceed $10^5$, which is more gain than needed in most applications. However, the frequency at which the op-amp gain begins to roll off, known as the bandwidth of the op-amp, may be considerably less than that required by the system. For example, in a linear amplifier, it is important that the bandwidth exceed the highest frequency of the desired input signal. If it does not, the frequency components of the input above the bandwidth cutoff will not be amplified by as much as those below. Consequently, the output will not be the scaled version of the input signal desired for a linear amplifier.

In this exercise, you will examine a method to use feedback in order to trade some of the surplus gain of an amplifier in exchange for a greater bandwidth. The following first-order system function is a simple model for an op-amp:

$$H(s) = \frac{Gc}{s+c}. \tag{11.11}$$

Here, $G = 10^5$ and $c = 2\pi(200)$. You will also simulate this system to see the importance of having the amplifier's bandwidth exceed the highest frequency component in the input signal.

### Basic Problems

(a). Analytically determine the open-loop DC ($s = 0$) gain for the system function given in Eq. (??), as well as the time constant of the system, i.e., the time at which $h(t) = h(0)/e$.

(b). Define the bandwidth of the system to be the frequency at which the magnitude of the frequency response is $1/\sqrt{2}$ times the DC gain, i.e., $H(0)/\sqrt{2}$. Analytically determine the bandwidth of the open-loop system.

(c). Define bH and aH to represent the coefficients of the numerator and denominator of $H(s)$ in Eq. (??). Use `freqs` and `impulse` to generate plots of the frequency response and impulse response of the open-loop system and verify your answers to Parts ?? and ??. The impulse response of a first-order system is negligible after roughly 5 time constants.

(d). Define `t=[0:3999]/8000` as the time index for one-half of a second sampled at $\Delta t = 1/8000$. Define the vector `x` to represent the signal

$$x(t) = \sin(2\pi(200)t) + \sin(2\pi(1000)t) \tag{11.12}$$

at the samples in `t`. Use `lsim` with `bH` and `aH` to simulate the result of amplifying $x(t)$ with an open-loop amplifier with the system function $H(s)$ in Eq. (??) and store the result in `y1`. Plot `x` and `y1` for the last 500 samples of the interval simulated. Is the output of the open-loop system a close approximation to a scaled version of the input?

## Intermediate Problems

The feedback system shown in Figure ?? can be used to increase the bandwidth of the amplifier.

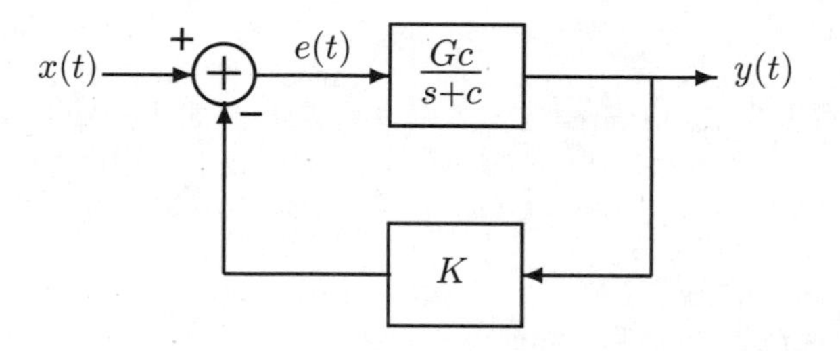


**Figure 11.4.** Feedback system to increase amplifier bandwidth.

(e). Analytically determine $Q(s)$, the system function for the closed-loop system in Figure ??. The function `feedback` can be used to check your answer. Specifically, if you define `K` to be the feedback gain, you can type

```
>> [bQ,aQ] = feedback(bH,aH,K,1,-1);
```

to compute the polynomials for the numerator and denominator of $Q(s)$ for that value of `K`. Use `feedback` with `K` $= 1$ and 2, and confirm that it returns the same coefficients as predicted by your analytic expression for $Q(s)$. Also, check that `feedback` returns the open-loop system function for $K = 0$.

(f). Analytically determine the DC gain, bandwidth, and time constant for the closed-loop system $Q(s)$. How do these quantities compare to those you found for the open-loop system?

(g). What value of $K$ will result in a closed-loop bandwidth that is five times as large as the open-loop bandwidth? What will be the effect on the DC gain and time constant of the system? Use this value for `K` with `feedback` to obtain `bQ` and `aQ` for the closed-loop system. Using these coefficients of the closed-loop system function with `impulse` and `freqs`, confirm that the DC gain, bandwidth and time constant of the closed-loop system have changed as you predicted they would.

(h). For the time samples in `t`, use `bQ` and `aQ` with `lsim` to simulate the effect of amplifying $x(t)$ from Eq. (??) with the closed-loop system. Plot the last 500 samples of the interval simulated. Is the output of the closed-loop system a closer approximation to a scaled version of the input than the output of the open-loop system was?

# Bibliography

This bibliography contains some of the more popular undergraduate level texts used for signals and systems. This list is not meant to be exhaustive, but should supply the reader with a number of choices for obtaining the relevant theoretical background needed to complete the exercises in this book. In addition, some mathematics texts are provided to allow the reader to pursue a more rigorous analysis of signals and systems theory.

## Signals and Systems

GLISSON, T. H., *Introduction to System Analysis.* New York, NY: McGraw-Hill, 1985.

JACKSON, L. B., *Signals, Systems, and Transforms.* Reading, MA: Addison-Wesley, 1991.

KAMEN, E., *Introduction to Signals and Systems.* New York, NY: Macmillan, 1987.

LATHI, B. P., *Linear Systems and Signals.* Carmichael, CA: Berkeley-Cambridge Press, 1992.

LUENBERGER, D. G., *Introduction to Dynamic Systems: Theory, Models, and Applications.* New York, NY: John Wiley, 1979.

MAYAN, R. J., *Discrete-time and Continuous-time Linear Systems.* Reading, MA: Addison-Wesley, 1984.

MCGILLEM, C. D. AND COOPER, G. R., *Continuous and Discrete Signal and System Analysis.* 3rd ed. New York, NY: Holt, Rinehart and Winston, 1991.

OPPENHEIM, A. V. AND WILLSKY, A. S. WITH NAWAB, *Signals and Systems.* 2nd ed. Upper Saddle River, NJ: Prentice Hall, 1997.

PAPOULIS, A., *Signal Analysis.* New York, NY: McGraw-Hill, 1977.

SIEBERT, W. M., *Circuits, Signals, and Systems.* Cambridge, MA: The MIT Press, 1986.

SOLIMAN, S. AND SRINATH, M., *Continuous and Discrete Signals and Systems.* New York, NY: Prentice Hall, 1990.

TAYLOR, F. J., *Principles of Signals and Systems.* McGraw-Hill Series in Electrical and Computer Engineering. New York, NY: McGraw-Hill, 1994.

ZIEMER, R. E., TRANTER, W. H., AND FANNIN, D. R., *Signals and Systems: Continuous and Discrete.* 2nd ed. New York, NY: Macmillan, 1989.

## Discrete-Time Signal Processing

BURRUS, C. S., MCCLELLAN, J. H., OPPENHEIM, A. V., PARKS, T. W., SCHAFER, R. W., AND SCHUESSLER, H. W., *Computer-Based Exercises for Signal Processing Using* MATLAB. Englewood Cliffs, NJ: Prentice Hall, Inc., 1994.

DUDGEON, D. E. AND MERSEREAU, R. M., *Multidimensional Digital Signal Processing.* Englewood Cliffs, NJ: Prentice Hall, Inc., 1984.

HAYKIN, S., *Adaptive Filter Theory.* Englewood Cliffs, NJ: Prentice Hall, Inc., 1991.

JOHNSON, D. E., *Introduction to Filter Theory.* Englewood Cliffs, NJ: Prentice Hall, 1976.

LIM, J. S., *Two-Dimensional Signal and Image Processing.* Englewood Cliffs, NJ: Prentice Hall, Inc., 1990.

OPPENHEIM, A. V., SCHAFER, R. W., AND BUCK, J.R., *Discrete-Time Signal Processing.* 2nd ed. Englewood Cliffs, NJ: Prentice Hall, 1999.

PARKS, T. W. AND BURRUS, C. S., *Digital Filter Design.* New York, NY: John Wiley, 1987.

RABINER, L. R. AND GOLD, B., *Theory and Application of Digital Signal Processing.* Englewood Cliffs, NJ: Prentice Hall, 1975.

## Mathematics Used in System Theory

CHURCHILL, R. V. AND BROWN, J. W., *Complex Variables and Applications.* 5th ed. New York, NY: McGraw-Hill, 1990.

GOLUB, G. H. AND VAN LOAN, C. F., *Matrix Computations.* 2nd ed. Baltimore: The Johns Hopkins University Press, 1989.

LIGHTHILL, M. J., *Introduction to Fourier Analysis and Generalized Functions.* New York, NY: Cambridge University Press, 1962.

MITCHELL, A. R. AND GRIFFITHS, D. F., *The Finite Difference Method in Partial Differential Equations.* New York: Wiley and Sons, 1980.

SIMMONS, G. F., *Differential Equations: With Applications and Historical Notes.* New York, NY: McGraw-Hill, 1972.

STRANG, G., *Introduction to Linear Algebra.* Wellesley, MA: Wellesley-Cambridge Press, 1993.

# Index